Wheat Marketing in Transition

ENVIRONMENT & POLICY

VOLUME 53

For further volumes:
http://www.springer.com/series/5921

Linda Courtenay Botterill

Wheat Marketing in Transition

The Transformation of the Australian Wheat Board

Prof. Linda Courtenay Botterill
Faculty of Business and Government
University of Canberra
Canberra, ACT 2601
Australia
Linda.Botterill@canberra.edu.au

ISSN 1383-5130
ISBN 978-94-007-2803-5 e-ISBN 978-94-007-2804-2
DOI 10.1007/978-94-007-2804-2
Springer Dordrecht Heidelberg London New York

Library of Congress Control Number: 2011944272

Printed on acid-free paper

Springer is part of Springer Science+Business Media (www.springer.com)

Preface

The origins of this book lie in the public furore that surrounded revelations in 2005 that Australia's monopoly wheat exporter, AWB Limited, had been one of the worst offenders in cooperating with Saddam Hussein's regime to bypass the sanctions imposed by the United Nations against Iraq. AWB Limited had been responsible for the payment of some $US221 million in kickbacks under the humanitarian component of the 'Oil-for-Food' program, which began in 1996 and ran until the invasion of Iraq by the Coalition of the Willing in 2003. In Australia, the involvement of AWB Limited became a major political scandal, threatening to end the careers of senior Cabinet Ministers in the Howard Government and, incidentally, providing a springboard for the rise of Kevin Rudd to the leadership of the opposition Australian Labor Party and ultimately to the Prime Ministership.

The present project was motivated by my strong sense that the reporting of the scandal, and the political debate around AWB Limited's actions, was incomplete. It was flawed in several important ways. First, the analysis was ahistorical; it was ignorant of the background to the monopoly wheat exporter's dealings with Iraq and its unhappy experience with earlier sanctions against Iraq. Second, it was confused about the nature of the company at the heart of the story. AWB Limited had its origins as the statutory Australian Wheat Board which had been established in 1948 but which underwent a process of privatisation in the mid-1990s. The privatised entity continued to be the beneficiary of a legislated monopoly over wheat exports, however, it was not a government agency and its employees were not government employees at the time the sanction-busting activities took place. Throughout the public debate in the mid-2000s, this important distinction was lost as media commentators and politicians continued to refer to AWB Limited as the Australian Wheat Board. Third, the debate did not recognise the essentially secretive nature of the international wheat trade; a trade which is dominated by several large, privately-owned companies who play their cards very close to their chests. The main implication of this characteristic of the wheat trade was the unrealistic expectation that officers within the Department of Foreign Affairs and Trade who were involved in the management of the sanctions could know whether the price of wheat on a contract had been inflated to disguise the payment of kickbacks.

This book seeks to provide a more considered and complete examination of the Oil-for-Food scandal; but that is only part of the story. The Australian Wheat Board was a longstanding and well respected institution of Australian rural policy. It outlasted many of the Australian Government's support measures for agriculture which were steadily dismantled from the 1970s onwards. The Wheat Board enjoyed strong support from wheatgrowers and it represented important collective values with their origins in the 1930s. The Australian Wheat Board, and then AWB Limited, was the institutional embodiment of these values and it is the institution of collective marketing on which this book focuses.

In the mid-1990s, I was employed as Manager, Strategic Planning at the Grains Council of Australia, the peak industry body representing Australia's graingrowers. In that role I was involved in the early stages of the debate around the privatisation of the statutory Australian Wheat Board, drafting discussion papers and attending meetings with growers to discuss the future of the collective marketing arrangements. I was exposed to the passionate attachment many growers felt for the Board and its role as monopoly exporter. I was also aware of the many challenges that these collective arrangements were facing in a policy environment which favoured deregulation and minimal government intervention in markets. The Oil-for-Food scandal probably accelerated the demise of collective wheat marketing in Australia but it was arguably inevitable. As the book describes, the industry underwent major changes in the 1980s, including the loss of the Wheat Board's monopoly over the domestic wheat market and the writing was on the wall for the remainder of the Board's statutory powers.

The story of the privatisation of the Australian Wheat Board has not been told and, as indicated above, the involvement of AWB Limited in the Oil-for-Food scandal has been given only an incomplete treatment to date. I hope that the account that follows will address this and, as well as filling an important gap in the history of Australian rural policy, will provide lessons for similar arrangements, notably the Canadian Wheat Board. The story of the Wheat Board highlights the pitfalls of privatising a monopoly in a way which left important powers with the new body and blurred the distinction between the government agency and a private company seeking to maximise its profits and its share price. As an account of a longstanding institution from cradle to grave, the book also seeks to fill a gap in the historical institutionalist literature which to date has been focused on institutional birth and survival. I hope that considering the demise of an institution and the factors which contributed to that end provide some insights into institutional strategies of reproduction; and the need to consider the unintended consequences of apparently successful adaptation to change.

I would like to thank the many grains industry insiders who spoke to me both on and off the record during my research for this book. I am also grateful to the former Chief Operating Officer of the Grains Council of Australia, David Ginns for releasing many of the Council's internal papers to the Noel Butlin Archives at the Australian National University; where they were not only accessible to me but are available to any other researchers interested in this fascinating piece of rural policy history. There are still many stories buried in those papers waiting to be told.

There are of course many other people to thank for supporting me as this project unfolded. My husband and dive buddy Bob is my escape from academia and our time spent together underwater provides a blissful and precious change of pace and focus. However, this book is dedicated to my father who, as the most over-qualified research assistant possible, made an invaluable contribution to this project at a time when I was feeling slightly overwhelmed and struggling to make progress. He has shown immense interest in my academic career and is probably unaware how much that means to me.

Linda Courtenay Botterill

Contents

1 Introduction ... 1
Analytical Framework ... 2
Historical Institutionalism ... 3
The Role of Values ... 6
Making Rural Policy in Australia ... 8
The Socio-Political Context ... 10
The Book's Structure ... 14
References ... 17

2 Australian Wheat Industry Policy in Context ... 21
Wheat in the Australian Economy ... 22
Wheat Exports ... 26
The National Contribution ... 28
The Rural Policy Paradigm ... 29
References ... 31

3 The Birth of Collective Wheat Marketing ... 33
Institutional Development and Change in the Wheat Industry ... 34
The Nature of the Policy Community ... 39
The Birth of the Australian Wheat Board ... 41
A Parallel Case: The Establishment of the Canadian Wheat Board ... 44
The International Context ... 46
Conclusion ... 48
References ... 49

4 From Orderly Marketing to Deregulation 1948–1988 ... 51
A Changing Rural Policy Paradigm ... 52
Institutional Development and Change ... 56
The Industries Assistance Commission Reports ... 57
The McColl Royal Commission ... 59
Reviews of Statutory Marketing ... 61

The Changing Face of the Rural Policy Community ... 63
Conclusion ... 65
References ... 65

5 **From Domestic Deregulation to Privatisation** ... 67
Institutional Change and Adaptation ... 69
Some Context: The National Competition Policy ... 72
The Privatisation Process ... 73
The First Attempt at Change: The Newco Debate ... 73
Hastening Slowly: The Second Debate ... 75
Government Policy and the Privatisation Process ... 83
Conclusion ... 85
References ... 87

6 **The Monopoly Wheat Exporter and the Dictator** ... 91
Iraq, Sanctions and Australian Wheat Exports ... 91
The UN Sanctions Regime and the Oil-for-Food Program ... 91
The Problem of Sanctions Implementation ... 98
Parallel Timelines ... 101
Oil-for-Food, AWB Limited and Institutional Reproduction ... 102
Conclusion ... 105
References ... 106

7 **The Aftermath of Oil-for-Food and the Death of an Institution** ... 107
The End of the Single Desk ... 109
The End of Grower Control ... 112
The Aftermath ... 113
The Values Dimension ... 113
Impact on the Policy Community ... 117
Conclusion ... 123
References ... 124

8 **Lessons and Reflections** ... 127
The Role of the Individual in the Demise of the Institution ... 129
Values, Politics and Institutions ... 130
Some Policy Lessons from the Death of Collective Wheat Marketing ... 134
How Not to Privatise a Monopoly ... 135
The Problem of Sanctions Implementation ... 137
Collective Wheat Marketing in Australia ... 139
References ... 140

Index ... 143

Abbreviations

ANZAC	Australia New Zealand Army Corp
ASW	Australian Standard White
ASX	Australian Stock Exchange
AWB	Australian Wheat Board
AWB Limited	The privatised Australian Wheat Board
AWBI	AWB International Limited—a subsidiary of AWB Limited
AWF	Australian Wheatgrowers Federation
CWB	Canadian Wheat Board
DFAT	Department of Foreign Affairs and Trade
DPIE	Department of Primary Industries and Energy
EEC	European Economic Community
EU	European Union
FAQ	Fair Average Quality
GATT	General Agreement on Tariffs and Trade
GCA	Grains Council of Australia (successor organisation to the AWF)
GMP	Guaranteed Minimum Price
IAC	Industries Assistance Commission
ISCWT	Iraqi State Company for Water Transport
IWA	International Wheat Agreement
MP	Member of Parliament
NFF	National Farmers Federation
NSW	The state of New South Wales
OFF	Oil-for-Food
SMA	Statutory Marketing Authority
SPU	Strategic Planning Unit
STE	Statutory Trading Enterprise
UAP	United Australia Party
UN	United Nations
UK	United Kingdom

US	United States (of America)
WEA	Wheat Export Authority (until 2008)
	Wheat Exports Australia (from 2008)
WIF	Wheat Industry Fund
WTO	World Trade Organisation

Chapter 1
Introduction

Keywords Policy process • Historical institutionalism • Values • Australia • Rural policy

There is something about wheat. Of all the basic foodstuffs, this grain seems to arouse the most passion and attract the greatest attention of governments. Works on the wheat industry frequently cite Socrates' observation that 'Nobody is qualified to become a statesman who is entirely ignorant of the problem of wheat' (see for example Dunsdorfs 1956: 263; Morriss 1987: 1). The market for grain has also been ranked with oil cartels and financial markets as a source of power and influence (Morgan 1979: 228). This book describes how one government, that of Australia, has dealt with the 'problem of wheat'. In so doing it tracks the life and death of collective wheat marketing as an institution, and its organisational manifestation in the form of the Australian Wheat Board, which was central to Australian wheat marketing for nearly 60 years. The manner in which the Board continued successfully to reproduce itself over time provides a neat example of how competing values play out in the policy process and how institutions having been established to further one set of values can effectively exclude other values from policy consideration.

The statutory Australian Wheat Board was established in 1948 following nearly two decades of debate about the collective marketing of wheat. Similar debates were under way in Canada which also opted for a statutory wheat marketing arrangement; the Canadian Wheat Board was established in 1935. Experience with a free market in the 1920s and 1930s had left Australian wheat growers very suspicious of 'middle men', and that distrust persisted well into the early years of the twenty-first century. The Australian Wheat Board embodied a set of broadly agrarian ideals of collectivism and grower control of wheat marketing. Over time, wheat marketing legislation in Australia was amended in response to changing economic conditions and evolving attitudes towards the role of government. In comparison with other areas of the Australian economy, and particularly with other areas of rural policy,

L.C. Botterill, *Wheat Marketing in Transition: The Transformation of the Australian Wheat Board*, Environment & Policy 53, DOI 10.1007/978-94-007-2804-2_1,

collective marketing of wheat remained remarkably persistent, both as an idea and as a policy approach. Central to this persistence was the Australian Wheat Board and its strong support structure in grower organisations and among wheatgrowers more broadly.

Wheat marketing has generally been treated as an arcane area of Australian public policy, of interest only to wheatgrowers, their representative organisations and the relevant government agencies. The statutory marketing arrangements in Australia, particularly the existence of the export monopoly or 'single desk', have attracted the attention of other governments and their grains industries, notably US Wheat Associates. Within Australia, however, wheat industry policy did not have a high public profile until 2003 when *The Australian* newspaper broke the story of the sanction-busting activities in Iraq by Australia's monopoly wheat exporter, and institutional successor to the Australian Wheat Board, AWB Limited. This brought wheat policy on to the breakfast tables of the general public and the unfolding saga provided nightly stories on the television news for months.

This book discusses the scandal surrounding the UN's 'Oil-for-Food' scheme which catapulted wheat marketing on to the Australian public agenda but it is not the whole focus of the study. The scandal raises some interesting questions about institutional evolution and change: how could a venerable institution so quickly become a national disgrace? How could it behave in such a way that it threatened the core values on which it was based? Why did wheatgrowers continue to defend the institution's actions long after allegations of kickbacks had been proven? This study goes back to the beginning and examines the institution of collective wheat marketing in Australia over its whole history. In so doing, it seeks to contribute to our understanding of the role of institutions in political life, the role of values in public policy, and the influence of policy players in defending their values against competitors in the policy space.

Analytical Framework

The story of the transformation of wheat marketing in Australia is framed in this book as an 'analytical narrative', that is, a case study which is embedded in an analytical model (Shepsle 2006: 34–35). The overall model which informs consideration of the case of collective wheat marketing is that of historical institutionalism, an empirically rich tradition which recognises the importance of institutions in political life and provides approaches to understanding their endurance. Although situated within this literature, the study departs from it in three important respects.

First, while it follows the literature in considering the creation and reproduction of the institution being studied, it also develops an earlier extension of the concept to include consideration of organisational death (Botterill 2011). This is not a huge departure from the work of others as institutional demise has been hinted at, but it has not been incorporated as a major part of the research. Much of the historical institutionalist literature has focused on institutional reproduction. Scholars in the field have pointed to the problem of unintended consequences of policy action but have not sustained this analysis to its logical conclusion. A snapshot approach to assessing strategies of reproduction does not allow for the unintended consequences of apparent success. Collective wheat marketing in Australia provides an ideal case study for the consideration of institutional birth, reproduction, and death, in which apparently successful adaptation to change contributed to the ultimate demise of the institution itself.

The second shift is in the present work's focus on values. The historical institutionalists have given attention to the embodiment of *ideas* in institutions but it is suggested that it is at the more fundamental level of *values* that fruitful work can be done on the role of institutions. Peters raises values in his study of institutionalism, suggesting that one of the core features of any institutional approach should be 'some sense of shared values and meanings among the members of the institution' (Peters 1999: 18). This actually moves beyond the ideas-focus of much of the literature and is consistent with the approach taken here, which draws on the work of public policy scholars on the role of values in the policy process.

Finally, the study borrows from the British variant of the policy network literature to consider how the policy context, within which the wheat marketing arrangements were sustained, remained uncontested while other sectors of the Australian economy were undergoing significant structural adjustment as a result of a changing economic paradigm. There is considerable overlap between these three analytic lenses. Values are essential to the understanding of cohesive policy communities and such communities can be sufficiently invested in particular institutions that they become important champions of institutional survival when there is pressure for change.

Historical Institutionalism

Historical institutionalism is the earliest of the 'new institutionalisms' which have arisen since the 1980s. Revived interest in institutions as an explanatory factor in political life was partly a response to perceptions of an excessive individualism that had developed in the political science discipline in the form of behaviouralism in the 1950s and 1960s, followed by the rise of rational choice (see for example Clemens and Cook 1999; Peters and Pierre 1998; Shepsle 1989). The institutionalist approach focuses on the role of institutions in structuring political life. There are some writers who dispute the incompatibility of individualist and institutional approaches (e.g. Dowding 1994) and the literature grapples with the issue of the role of agents within institutional structures (see for example Hall and Taylor 1998; Hay and Wincott 1998). This study recognises the importance of action by individuals within an institution, as illustrated in the Oil-for-Food scandal. However, the analysis leans towards the institutionalist position that, while actors clearly have free will and can act against the interests of their institutions, their activities are constrained by their institutional context.

As Streeck and Thelen observe 'Definitions of institutions abound' (2005: 9). It is therefore important to clarify that the institution under discussion in this book is the system of collective wheat marketing, first enacted in Australia in the *Wheat Stabilization Act 1948* and then continued in various iterations of the *Wheat Marketing Act 1989*. This institution was then given organisational form through the statutory Australian Wheat Board. The core features of collective marketing, the 'pooling' of the export wheat crop, and an export monopoly with legislative backing, were retained by the company AWB Limited after the Wheat Board was privatised.

Historical institutionalists argue that institutions need to be seen in historical context. They reject the functionalist notion that all extant institutions are in place because they serve a purpose *today*, arguing that institutions are developed in a particular context and embody a particular political compromise that was relevant at the time of their creation. Institutions then continue to exist through various strategies of reproduction and also as a result of path dependence, a central concept in the historical institutionalist literature. In essence, path dependence suggests that decisions made at the time of the creation of an institution constrain the options available in the future. This suggests that the creation of an institution is an important 'win' for political forces seeking to promote their interests; once a political compromise is embodied in an institution it is much harder to revisit the original debate and consider alternative policy options. As Pierson (2004: 21) points out

> the probability of further steps along the same path increases with each move down that path. This is because the *relative* benefits of the current activity compared with once-possible options increase over time.

The path dependence idea has attracted criticism for two key reasons. First, while it is very useful in explaining stability, it does not account for institutional change. Institutions may embody a set of values long after those values are salient, but they do not remain frozen in time and, in order to survive, adopt various strategies of reproduction. The second major criticism relates to the early tendency in the literature to equate path dependence with path *determinacy*. Empirical work by historical institutionalists, however, has not borne out the argument, put explicitly by Mahoney (2000: 511) and implied by others, that path dependent events are 'relatively deterministic'.

Institutional change was originally addressed in the literature with the fairly crude notion of 'critical junctures' and 'punctuated equilibria' which involved some form of exogenous shock to the institution, initiating institutional change. As Peters (1999: 68) notes, this is not particularly useful as an analytical concept as it provides 'little or no capacity to predict change'. It also denies the capacity for actors to transform the institution from within through adaptation and learning. The present case study provides a clear example of institutional transformation brought about by a combination of external pressure and endogenous change. More recent theorising has addressed this critique to some extent through discussion of the potential transformative effects of incremental change and through the development of the concepts of displacement, layering, drift, conversion and exhaustion (Streeck and Thelen 2005). These understandings of forms of change leave room for the actor within the institution to respond to a shifting political and economic environment in a manner which is not as drastic as that implied by the critical juncture explanation for change. For example, the concept of layering suggests that 'reformers learn to work around those elements of an institution that have become unchangeable' (Streeck and Thelen 2005: 23) while making other changes to modernise the institution and keep it relevant.

Institutions are sustained not just by the phenomenon of path dependence and the fact that they are too costly to remove. They also benefit from the development of complementary institutions that provide a support network for their activities and which are invested to varying degrees in the institution's survival. As discussed in subsequent chapters, the body representing Australia's graingrowers, the Grains Council of Australia, and its predecessor body the Australian Wheatgrowers Federation, were heavily invested in the collective wheat marketing arrangements to the point that the demise of these arrangements was followed very quickly by the virtual collapse of the Grains Council as the recognised representative body for the Australian grains industry.

Mahoney (2000) proposes a fourfold typology for explaining institutional reproduction: utilitarian, functional, power and legitimation. The utilitarian explanation proposes that institutions continue to exist because 'any potential benefits of transformation are outweighed by the costs' (Mahoney 2000: 517). The functional explanation situates the role of the institution within the overall functional needs of the political system. Elite actors are central to the power explanation; an institution with powerful supporters will remain relevant. In this explanation, institutional change results from shifts in the societal balance of power away from the original political compromise embodied in the institution. The final explanation relates to the legitimacy of the institution and this element of Mahoney's typology is particularly relevant to the story of the Australian Wheat Board. It draws attention to the role of values in the survival of an institution and also, as will be discussed in Chap. 7, alerts us to the prospect that an institution may rely on one set of values for its legitimacy while pursuing a completely different set of values internally. In the case of collective wheat marketing in Australia, belief by growers in the export monopoly and the values embedded within it was an important factor in the fight over the institution's future. As the Oil-for-Food scandal unfolded, AWB Limited continued to appeal to the values that underpinned the legitimacy of the monopoly arrangements, even as it became increasingly apparent that the employees involved in the scandal were driven by an entirely different set of values.

As critics of the literature have pointed out, historical institutionalism tends to focus either on 'the *construction, maintenance* and *adaptation* of institutions' (Sanders 2006: 42) or, even more narrowly, either their creation (Hay 2006: 60) or their 'persistence' (Peters 1999: 67). This limits the analysis. Conclusions drawn about successful strategies of reproduction at a single point in time may prove premature when considered in the context of the institution's full life cycle.

Pierson has drawn attention to the need to consider 'dynamic processes' when examining an institution's survival; he calls for work that 'carefully investigates processes unfolding over time' (Pierson 2000: 494). An emphasis on path dependence and unintended consequences implies the rejection of a 'snap shot' view of institutional development and change but this cannot be done thoroughly while the process of development continues. Just as determining the start date for path dependent processes may be problematic (Mahoney 2000: 511), determining an end point for analysis can similarly run the risk of missing key events and outcomes. This is what makes a case like collective wheat marketing in Australia so valuable as an application of the historical institutionalist approach. It has an identifiable beginning and an end, and as such provides an opportunity to identify adaptations to change which were successful in the short term but over the long term proved counterproductive.

The Role of Values

The need for change and how it arises draws attention to an important feature of institutions; they are an embodiment of political compromises and the values these represent. Historical institutionalism has drawn attention to the role of *ideas* in the creation and ongoing existence of institutions. It is probable that the term 'ideas' is being used in the literature synonymously with 'values' but there is a difference of degree. 'Value' implies something more deeply held than an idea, which can be transient and discarded when replaced with a superior notion. The institution of collective wheat marketing described in this book is based on something more fundamental than an idea; it is anchored in deeply held values which have a normative force. The significance of identifying core values as opposed to simply 'ideas' can be illustrated by the work of Sabatier (1988) who argued that advocacy coalitions within the policy process operate with three levels of beliefs: deep core values, near or 'policy core' values, and secondary aspects. Sabatier describes changes in deep core values as 'akin to religious conversion' (144); they are very resistant to change and policy actors will accordingly act to protect their deep core beliefs as these represent fundamental understandings about the way the world operates. As discussed in later chapters, this almost religious characteristic of deeply held values is evident in the fervour with which many of the supporters of collective marketing reacted to the prospect first of deregulation of the domestic wheat market and then to the end of the export monopoly.

The proposition that values are central to the policy process is not new. David Easton pointed to the essential values-based nature of politics in 1953 with his observation that 'politics is the authoritative allocation of values' (Easton 1953: 129). Lindblom (1965: 227) similarly argued that 'The question of how values are weighted into decisions or resultant states of affairs is central to the study of public decision making, because government can be regarded in large part as machinery for resolving value conflicts'. All political communities comprise groups and individuals with different normative views of the world and these positions inevitably conflict. A successful polity manages these conflicts by 'balancing' the different perspectives in a manner which broadly satisfies its members. Policies are developed which deliver sufficient comfort to each group that the legitimacy of the overall system is accepted. A typical trade-off within a liberal democracy itself is that between the liberal instinct for personal freedom, unencumbered by government intervention, and the democratic tendency towards the collective good. This plays out in debates over checks and balances in political structures with different political systems arriving at different compromises. In policy terms, value conflicts are evident, for example, in economic policies which determine the division of wealth in the community between labour and capital, in debates over the environmental impact of economic activity, in the trade-off in some policies between equity and efficiency, and in the debates over the loss of some individual liberties in the response to terrorism.

Scholars have long been interested in how these value conflicts are resolved in the policy process. Charles Lindblom identified this element of the policy process in his influential 1959 article 'The Science of "Muddling Through"' which challenged prevailing ideas that the policy process was a rational one of setting goals, identifying available policy options, ranking them and then choosing the best available policy. Lindblom argued that policy in the real world is made incrementally; there are important resource constraints on policy makers that prevent their accessing all available information and undertaking a 'rational comprehensive' process. There is little disagreement in the literature that, as an empirical observation of the policy process, Lindblom's description of incrementalism was right. More controversially, he suggested that incrementalism was also the best way to make policy because of the way in which it managed value conflicts. Any policy decision privileges one value set over another; the decision to 'fight inflation first' suggests stability in capital markets is more important than full employment. Through incremental policy making, values that have been overlooked in earlier iterations can be addressed if the consequences of their neglect become acute. Lindblom described the policy process as both serial and remedial. Policies are frequently modified by government as implementation reveals unintended outcomes, and as policy learning occurs. Lindblom also argued that some values are intentionally overlooked because policy makers know that policy is not made for all time and will be revisited. Policy makers know that their information is incomplete, that circumstances will change and that they will have overlooked an element of the policy issue which will need addressing in the future.

There has been a recent growth in the literature on the role of values in the policy process and the strategies adopted by policy makers in dealing with value conflict. Thacher and Rein have proposed three strategies: 'policy cycling', firewalls and casuistry. Policy cycling is very similar to Lindblom's incrementalism, in that policy makers 'focus on each value sequentially, emphasizing one value until the destructive consequences for others become too severe to ignore' (Thacher and Rein 2004: 463). The firewall strategy is more of an avoidance strategy than a balancing act. Under this model, policy-makers quarantine incompatible or conflicting values by addressing them separately, often in different organisations. This approach ensures that 'each value has a vigorous champion' (Thacher and Rein 2004: 463) in the policy process but can also lead to anomalies such as governments campaigning against smoking on health grounds while agricultural policy supported an inefficient industry protected from market forces (IC 1994), as was the case in Australia for a period in the 1980s when 'tobacco leaf growing was the most heavily government-subsidised economic sector in Australia' (Studlar 2005: 256). Firewalls can also be used to exclude particular issues from policy debate altogether. For a long time in Australia there has been an elite consensus which has kept the issue of the death penalty off the policy agenda. The final strategy identified by Thacher and Rein they label 'casuistry', by which they mean that policy makers address value conflicts on a case by case basis by analogy with similar circumstances.

In her study of value conflict and policy change, Stewart (2006) has built on Thacher and Rein's work to add hybridization, incrementalism and bias to the list of strategies. Stewart's concept of hybridization is similar to Streeck and Thelen's work on layering in that she sees new values co-existing, at times uncomfortably, with older approaches. This is not necessarily a practical response to value conflict but can 'satisfy the need for an all-embracing rhetoric' (Stewart 2006: 188). Lindblom's definition of incrementalism clearly encompasses the Thacher and Rein 'cycling' idea, but Stewart's distinguishes between a more conventional, linear understanding of incrementalism and a form of cycling which she attributes to Thacher and Rein that is more cyclical, involving 'flip-flops' and backlashes (2006: 192). Her concept of bias reflects Mahoney's focus in the area of institutions on power relations. The bias strategy for managing values defines issues 'in' and 'out' of consideration; this concept is explored further below in the consideration of the role of policy communities in the policy process. Stewart refers to dominant policy paradigms and the increasing 'technicization' (2006: 191) of the policy process as influential factors in ensuring one set of values dominates over others.

One important mechanism for preserving a particular set of values is to embed them within an institution. As the historical institutionalist literature illustrates, institutions are remarkably resilient and adopt various strategies for reproduction which prolong their existence, often after the values on which they were based have ceased to be relevant in a contemporary policy setting. An examination of the Australian Wheat Board and its adaptation to changing economic circumstances and evolving policy priorities suggests resilience. The case study tracks the establishment of an institution, its response to changing circumstances and its attempts to protect its core values while trading away lesser elements of its original charter.

Making Rural Policy in Australia

The third literature on which this study draws is the work on policy networks and policy communities. A notable feature of rural policy in Australia has been the small circle of interest groups and government agencies which are engaged in the debate over policy settings. The broader public, the media and the wider public sector generally do not regard rural policy as a priority area for attention and seem relaxed about leaving policy choice in the hands of a few rural policy specialists; until an issue such as the Oil-for-Food scandal grabs their attention. One way of conceptualising this closed arrangement is through the notion of 'policy communities' which emerged in the British literature on networks.

The importance of networks of influence was first identified, in the US context, by Hugh Heclo (1978) who rejected the then current understanding that policy making was dominated by an 'iron triangle' made up of congressional committees, government agencies and peak lobby groups. Heclo considered this picture to be too narrow and argued that there were in fact broader 'issue networks' of interested players who were involved in policy negotiation and who had varying levels of power and expertise in the relevant policy area (Heclo 1978: 102–105). About the same time that he was writing, Richardson and Jordan in the UK published their work on the 'post parliamentary democracy' (1979). They argued that policy change only happened with the agreement of the relevant policy community, comprising a restricted range of groups with access to the policy process. This concept was developed further by Marsh and Rhodes (1992: 251) who proposed a continuum between loose 'policy networks' and tighter 'policy communities'. The distinction between types of network was drawn on the basis of their membership, the level of integration between those members, the resources they hold and the distribution of power. The policy communities end of the spectrum is made up of groups who have shared values and common world views. Their membership is limited, there exists a resource exchange relationship and members interact regularly. There is a balance of power among members and the arrangement is highly stable. They have a shared understanding about what problems are important and what solutions are appealing and do-able. The conflicting values in the policy process are managed by excluding particular voices from the policy community. Agriculture is often cited in the literature as a policy area containing archetypal policy communities (Daugbjerg 1999; Grant and MacNamara 1995; Peterson and Bomberg 1999: 138–141; Smith 1990). Although the theoretical value of the policy communities concept has been subject to debate (Dowding 1995, 2001; Marsh and Smith 2001), it remains a useful heuristic device for understanding and interpreting the way policy is made in a particular policy area and can help explain the values 'bias' of which Stewart writes.

It is worth noting that discussion of networks has not been limited to the US and UK literature. The European scholarship on networks has, however, taken a different approach, focusing on the emergence of networks in the context of governance arrangements in which 'the democratic system is losing part of its political functions to other—more elusive—institutions in society' (Godfroij 1995: 182). In this context, network analysis is a component of the study of public management (Kickert 1997). For the purposes of this book, the British literature is the most relevant with its emphasis on a small group of highly knowledgeable and influential policy players with a shared understanding of the policy issues. There was clearly a policy community, in Marsh and Rhodes's terms, dominating and controlling policy for the Australian wheat industry for the best part of 60 years. It very successfully excluded opponents of the export monopoly from serious policy debate until the Cole Inquiry into the Oil-for-Food scandal fundamentally altered the policy making environment. The revelations before the Inquiry provided the type of external shock which is needed to break the hold of a closed policy community on the policy process. As Smith (1991) argued in his study of the food policy community in the UK, issues which break up the value consensus within a policy community can provide openings for new players and can result in policy change. As later chapters illustrate, only the most devout supporters of the export monopoly continued to defend unconditionally AWB Limited's actions in Iraq following the Cole Inquiry.

The Socio-Political Context

There are two key features of Australian politics and culture which provide important background to the story of collective wheat marketing in this country. The first relates to one of the apparent anomalies in the Australian political landscape; that is, the existence of an identifiable agrarian party in one of the most urbanised countries in the world. The National Party of Australia first entered the Commonwealth Parliament in 1920 as the Country Party. It is the second oldest of Australia's three major political parties and has been defying its critics and their predictions of the party's imminent demise since its creation. The earliest record of the latter is from 1921 (Davey 2006: 23). The National Party is important in the story of collective wheat marketing because the retention of monopoly marketing for wheat, and particularly the export monopoly arrangements, has been an article of faith for the party. This largely reflects the demographic within the rural community that votes for the National Party which has tended to be concentrated in areas of rural Australia characterised by the smaller, less market-oriented family farms rather than the bigger end of town. Australia still has a low level of penetration by agri-business with only a tiny percentage of farm businesses operating as incorporated entities. The category of 'larger-than-family' farms is important in terms of output, accounting for over 80% of farm product, but numerically it is the smaller operations that dominate.

At the 2007 Federal election, the National Party attracted only 5.5% of the primary vote (Woodward and Curtin 2009: 40) and it does not even attract the majority of votes from farmers (Bean 2009: 72). Its significance arises from its longstanding coalition arrangement with the Liberal Party of Australia and its predecessors. This coalition has confounded non-Australian contributors to the party politics literature resulting in the regular treatment of Australia as a stable two-party system (Botterill 2009). There is no doubt, in recent years, that the National Party has lost policy influence but it is still an identifiable party with its own values and its own constituency. With the notable recent exception of the 2008 Western Australian election, the Liberal-National Party coalition generally presents itself to the electorate as a single entity with a common policy platform; it is quite different from the coalitions, described in the largely Euro-centric party politics literature, which are negotiated after the polls close in order to construct a workable government. Following most elections in which the Coalition is victorious, the Liberal Party needs the National Party's seats in order to form a majority; and yet the National Party does not trade on this balance of power position to win particular concessions. There is little or no chance that the federal National Party would consider formal coalition with the Labor Party, although it has entered into some arrangements at State government level. In 2010, the National Party's strategy of coalition came under considerable public scrutiny when Australia faced the highly unusual situation of a hung Parliament as a result of a federal election held in August. Following the election, with neither the Labor Party nor the Liberal-National Party Coalition holding sufficient seats to form government, three rural independents, ironically all former members of the National Party, found themselves with the balance of power. In order to secure the independents' support, both the major parties entered negotiations which focused on improving outcomes for rural and regional Australia. The National Party was left on the sidelines watching others use the power that it had held on many occasions in the past, but had chosen not to exercise.

The National Party's significance to the wheat debate is that it has been prepared to fight its coalition partners on policy issues which are of importance to its core constituency. These fights have been strident and have frequently been over questions relating to the support for agricultural industries. Following the revelations of the inquiry into the Oil-for-Food scandal, clear differences emerged between the Coalition partners as representatives of the two parties traded insults over the future of the export monopoly for wheat. Even as late as 2010 when the wheat export monopoly was seemingly a lost cause, members of the National Party were passionately defending the need for a return to earlier industry arrangements.

The second contextual feature relates to the place of farmers in Australian culture. In common with other developed western nations, Australian culture and identity is influenced by a deeply held, but largely unacknowledged agrarian romanticism. This is reflected in popular culture, in a broad sympathy for farmers and the hardships they face and generally in a relaxed disregard for the intricacies of rural policy (Botterill 2006). Although Australian agriculture has an image of being productivist, efficient and export-oriented, there exists a strong element of agrarian sentiment associated with farming as an activity. Agriculture remains dominated by the family farm although, as noted above, there is an important group of 'larger-than-family' farms in operation. As a consequence, the images associated with small scale farming remain powerful and influence policy, both overtly and more subtly, as well as providing members of the policy community with a language with which they can confine debate and deflect criticism (Botterill 2006).

Agrarianism is a nebulous concept which has been adapted and reinterpreted over time. Montmarquet (1989) provides an insightful history into the evolution of the agrarian ideal, illustrating its many interpretations from Aristotle to the agrarian reformers of the twentieth century concerned with sustainability. In the Australian context, Waterhouse (2005) shows how the 'Vision Splendid' was reinterpreted from the radical tradition of the lone, itinerant swagman to the more traditional, morally virtuous, pioneering family. The common thread in these agrarian visions is that agriculture has a special place in developed economies. The gist of the argument is that, without agriculture, human societies would not have made the transition from hunter-gatherers to settled communities. Settlement facilitated specialisation and the evolution of art and culture; in a nutshell it gave birth to 'civilisation'. The connection between farming and nature also adds to its mystique; farmers have a hard life but their activities are basic, natural and, by implication, wholesome. As a result non-farming communities retain a level of admiration and sympathy for the family on the land. Heathcote expressed this sentiment, in the context of drought relief, as follows:

> In any catastrophe, public sympathy goes out to the victims, but when those victims are the sons of the soil, on the margins of the good earth, struggling to give us our daily bread, the emotional response is tremendous and objectivity is often left behind. (Heathcote 1973: 36)

The seminal definition of agrarianism was developed by Flinn and Johnson based on a survey of Wisconsin residents in 1971 and through the analysis of editorials published in farm journals between 1850 and 1969. They identified the following five 'tenets of agrarianism':

- '*farming is the basic occupation on which all other economic pursuits depend for raw materials and food*'
- '*agricultural life is the natural life for man; therefore, being natural, it is good, while city life is artificial and evil*'
- farming delivers the '*complete economic independence of the farmer*'
- '*the farmer should work hard to demonstrate his virtue, which is made possible only though* [sic] *an orderly society*'; and
- '*family farms have become indissolubly connected with American democracy*'. (Flinn and Johnson 1974: 189–194 – italics in original)

This reflects the sentiments described above; as Montmarquet (1989: VIII) describes it, 'the idea that agriculture and those whose occupation involves agriculture are especially important and valuable elements of society'. JS Mill and Thomas Jefferson subscribed to the belief that farming was conducive to desirable and moral behaviour and saw an inherent *social* value in the promotion of small scale agriculture. In Mill's case, he argued of small scale European peasant farming that

> no other existing state of agricultural economy has so beneficial an effect on the industry, the intelligence, the frugality, and prudence of the population … no existing state, therefore is on the whole so favourable both to their moral and physical welfare. (Mill 1893: 374)

Jefferson went further. In his vision for America

> agriculture was not primarily a source of wealth, but of human virtues and traits most congenial to popular self-government. It had a sociological rather than an economic value. This is the dominant note in all his writings on the subject. (Griswold 1946: 667)

Jefferson himself wrote that

> The loss by the transportation of commodities across the Atlantic will be made up in happiness and permanence of government. The mobs of great cities add just so much to the support of pure government, as sores do to the strength of the human body. It is the manners and spirit of a people which preserve a republic in vigor. (cited in Griswold 1946: 668)

This idea that agriculture is an essentially good undertaking, combined with the city-country dualism, is also found in the Australian variant of agrarianism, 'countrymindedness'. The term, of uncertain origin but traceable at least to the beginnings of the Country Party in the 1920s, has been defined by Aitkin (1985: 35) as follows:

(i) Australia depends on its primary producers for its high standards of living, for only those who produce a physical good add to a country's wealth.
(ii) Therefore all Australians, from city and country alike, should in their own interest support policies aimed at improving the position of primary industries.
(iii) Farming and grazing, and rural pursuits generally, are virtuous, ennobling and co operative; they bring out the best in people.
(iv) In contrast, city life is competitive and nasty, as well as parasitical.
(v) The characteristic Australian is a countryman, and the core elements of the national character come from the struggles of country people to tame their environment and make it productive. City people are much the same the world over.
(vi) For all these reasons, and others like defence, people should be encouraged to settle in the country, not in the city.
(vii) But power resides in the city, where politics is trapped in a sterile debate about classes. There has to be a separate political party for country people to articulate the true voice of the nation.

It is worth noting both points (iv) and (v). Australia is a highly urbanised society and yet our popular self-image frequently includes the 'bush' and the 'outback' as defining characteristics. Point (vii) reflects the basis for the establishment of the Country/National Party and its continuing appeal to rural voters. Recent opinion polling reveals that Australians generally continue to subscribe to these sentiments with a large majority of the population supporting the view that agricultural production and rural living are very important for Australia's future. Opinion is also generally more positive about the qualities that apply to rural residents than to their urban counterparts (McAllister 2009).

Against this backdrop, it is perhaps not surprising that agrarian values have an influence over policy at some level. Although they are rarely articulated explicitly, their influence can be identified in a range of policy settings. As with all areas of government deliberation, agricultural policy development can be understood as an exercise in balancing competing societal values.

In spite of their pervasiveness in popular culture, rural issues are generally poorly reported in the mainstream media and, when they are, stories tend to resort to stereotypes. Reporters (and many politicians) feel obliged to dress in moleskins and chambray or checked shirts when visiting rural areas, occasionally topped off with an Akubra, the ubiquitous Australian bush hat. A consequence of this superficial engagement with rural issues is that rural policy has been dominated, as it is elsewhere, by a closed policy community which operates largely independently and is based around an agreed set of values which translate into a common understanding of the nature of the policy problem and 'acceptable' policy solutions (see for example Botterill 2005; Smith 1990). This closed policy process has allowed particular values to dominate without the involvement of alternative perspectives. The generally uncritical public debate has enabled this policy approach and was an important factor in the process which led to the very odd privatisation of the

Australian Wheat Board which occurred in 1998 and which is described in Chap. 5. Not only was the process totally different from any other privatisation occurring at the time, it was also undertaken without reference to broader government policy and without the involvement of agencies usually central to such economic policy debates (Aulich and Botterill 2007).

The story of the collective marketing of wheat in Australia is therefore one of deeply held values, represented by an important political party, embodied in a long standing institution and left largely unchallenged for nearly 60 years.

The Book's Structure

The book is set out as follows. Chapter 2 provides some background information about the Australian wheat industry and early Australian rural policy. It discusses the evolution of the wheat industry, its spatial characteristics and its place in the Australian economy. This is provided in order to put the industry in its context as one of Australia's most important export industries and to put Australian wheat exports in their global context. It is also provided to assist non-Australian readers to understand more about the conditions under which the Australian wheat industry has developed.

Chapter 3 provides background on early debates around wheat industry policy and then discusses the birth of the Australian Wheat Board in 1948. The chapter draws parallels with the Canadian experience and also introduces the values that were at the heart of the development of collective marketing and the context within which the policy settings were developed. These values were very clearly shared within the international community and were reflected in the series of International Wheat Agreements that were negotiated around the time of the Wheat Board's creation.

Chapter 4 tracks important changes in the rural policy paradigm in Australia and examines the implications for collective wheat marketing. Until the 1970s, Australian rural policy was very similar to that of comparable developed countries. It was highly interventionist with a large range of policies aimed predominantly at stabilising the incomes of farmers but also at ensuring fair prices for consumers. In the 1930s and 1940s Australia experienced low farm incomes and 'outright poverty' (McKay 1972: 29) in many areas; however, by the 1950s the sector was booming. Agriculture was increasingly seen as a source of critical foreign income for the import of manufacturing inputs and other goods important to a growing economy. The nation's prosperity was dependent on rural exports and it almost literally 'rode on the sheep's back'.

By the 1960s, economists and other commentators were calling for a winding back of agricultural support programs which were seen as unsustainable. This deregulatory push inevitably had implications for the wheat industry which are also discussed in this chapter. The overview in Chap. 4 is briefer than the consideration later in the book. First, the evolution of the wheat marketing arrangements was very slow in its first four decades until 1989 when the policy hit a critical juncture with a government decision to remove a key plank of the policy – the Wheat Board's monopoly over the domestic market for wheat. Second, the early years of collective marketing in Australia have been well covered by Whitwell and Sydenham (1991) who finish their account in 1988.

Chapter 5 explores the first major challenge to the collective wheat marketing arrangements; the 1989 deregulation of the domestic wheat market. The consequences of this policy change were profound for the grains industry. The Australian Wheat Board responded to the introduction of competition on the Australian market by engaging employees with different skills; employees who were driven by more instrumental values than the collectivism of the traditional marketing staff. The Grains Council responded to the new policy environment by initiating a strategic planning process for the industry in an effort to stave off further unwanted government-driven change. Both of these responses were critical to the future of the collective marketing of Australian wheat.

Chapter 5 then describes the most fundamental change to wheat marketing arrangements since 1948 and the biggest challenge faced by the industry; the privatisation of the statutory Australian Wheat Board. The chapter tracks the debate from 1995 to 2001 as the industry debated the future of its key institution in the face of a changing international trade environment and important changes in government policy. The most influential sectors of the industry remained steadfastly attached to the collective marketing of the export wheat crop and were very successful in preserving the value at the very heart of the arrangements – the export monopoly.

Chapter 6 discusses the series of events which were the catalyst for the demise of the collective marketing arrangements; the scandal surrounding AWB Limited's[1] involvement in undermining the sanctions against Iraq through manipulation of the United Nations' Oil-for-Food program. Debate around the Oil-for-Food scandal has been very limited in its scope. It has focused largely on the role of the Australian Government and what it knew about the kickbacks and when it knew. The issue was politically significant as it drew attention to a member of the Opposition front bench, Kevin Rudd who, not long after his involvement in pursuing the government of the day over the scandal, took over as leader of the Australian Labor Party and went on to win the Federal election in 2007. The parliamentary debate about the scandal undoubtedly contributed to tarnishing the reputation of the incumbent Liberal-National Party government and also exposed important policy differences between members of this long standing coalition.

---

[1]One of the confusions surrounding much media reporting about the Oil-for-Food Program was that the entity which engaged in sanctions-busting behaviour was a private company, AWB Limited, not its predecessor, the statutory Australian Wheat Board. In order to avoid perpetuating that confusion, this book uses 'the Australian Wheat Board' or 'the Wheat Board' when referring to the statutory arrangement and 'AWB Limited' when referring to the privatised body. Avoiding the use of the acronym 'AWB' for the statutory body may be cumbersome in places but it is considered essential to making the clear distinction between the government body and the private company.

The debate was, however, narrow in two important respects. First, it did not consider whether AWB Limited's behaviour was in fact particularly extraordinary given the history of sanctions implementation and the complexity of the Oil-for-Food program. Consideration of why and how AWB Limited became involved in Saddam Hussein's 'gaming' of the sanctions regime (Meyer and Califano 2006, xix) did not address a fundamental problem of sanctions implementation; that is, the gap that so often exists between the State's commitment to sanctions and the establishment of the necessary monitoring and enforcement mechanisms behind domestic borders to ensure that the private actors who effectively are required to implement the sanctions actually do so (Botterill and McNaughton 2008). The second deficiency in the critique of the Oil-for-Food scandal was the assumption by many commentators that the intricacies of the international wheat trade are easily mastered and that middle ranking public servants in the Australian Department of Foreign Affairs and Trade should have been able to detect the kickbacks built into the prices on the contracts sent through for their scrutiny. This assumption is flawed for several reasons, not least of which is the complexity of international grain markets (Sewell 1992: xv) and the lack of transparency in their operation (Morgan 1979: 223, 224). Also of significance was the fact that AWB Limited had only recently been a part of government and a trusted source of intelligence on wheat markets (Botterill 2007).

Chapter 7 considers the aftermath of the Oil-for-Food scandal and the death of the institution of collective marketing arising from the government's response to the *Report of the Inquiry into certain Australian companies in relation to the UN Oil-for-Food Programme* by the Hon Terence Cole AO RFD QC. The Inquiry was protracted, taking much longer than the originally scheduled 3 months and guaranteeing that AWB Limited occupied the front pages of major daily newspapers for the best part of a year. The Coalition Government, which commissioned the Inquiry and received the report, was equivocal in its response. The Liberal Prime Minister hinted that the industry was facing the end of the statutory export monopoly, while his coalition partner the National Party argued that the arrangement was to be preserved. The decision was ultimately one for the incoming Labor government in 2007 which wasted little time in announcing its intention to end the monopoly (Burke 2008). This chapter explores the death throes of collective marketing and reflects on what this means for the values embodied in the institutional arrangements that underpinned wheat policy for nearly six decades.

Chapter 8 concludes the book by revisiting the key themes of the case study: the importance of values in the policy process and how they become embedded in institutions and protected by policy communities. As the institution of collective marketing entered its death throes, its champions continued to appeal to the values on which it was built; promoting the growers' interests and protecting them from exploitation by middle men. In the end the institution was not able to continue to reproduce itself in the face of changing economic and political conditions, and in response to the actions of its own employees whose values had departed significantly from those that provided legitimacy to the export monopoly. The chapter concludes with some lessons from the Australian experience on how not to privatise a monopoly and the pitfalls of sanctions implementation.

References

Aitkin D (1985) “Countrymindedness”—the spread of an idea. Aust Cult Hist 4:34–41

Aulich C, Botterill L (2007) A very peculiar privatisation: the end of the statutory Australian Wheat Board Australasian Political Studies Association Conference, Monash University, Melbourne, 24–26 September 2007

Bean C (2009) Trends in National Party support. In: Botterill LC, Cockfield G (eds) The National Party: prospects for the great survivors. Allen & Unwin, Sydney

Botterill LC (2005) Policy change and network termination: the role of farm groups in agricultural policy making in Australia. Aust J Polit Sci 40(2):1–13

Botterill LC (2006) Soap operas, cenotaphs and sacred cows: countrymindedness and rural policy debate in Australia. Public Policy 1(1):23–36

Botterill LC (2007) Doing it for the growers in Iraq?: the AWB, oil-for-food and the cole inquiry. Aust J Publ Admin 66(1):4–12

Botterill LC (2009) An agrarian party in a developed democracy. In: Botterill LC, Cockfield G (eds) The National Party: prospects for the great survivors. Allen & Unwin, Sydney

Botterill LC (2011) Life and death of an institution: the case of collective wheat marketing in Australia. Public Admin 89(2):629–643

Botterill LC, McNaughton A (2008) Laying the foundations for the wheat scandal: UN sanctions, private actors and the Cole inquiry. Aust J Polit Sci 43(4):583–598

Burke T, the Hon MP (2008) Government takes next steps to end wheat monopoly. Media Release by the Minister for Agriculture, Fisheries and Forestry DAFF08/006B, 6 February 2008

Clemens ES, Cook JM (1999) Politics and institutionalism: explaining durability and change. Annu Rev Sociol 25:441–466

Daugbjerg C (1999) Reforming the CAP: policy networks and broader institutional structures. J Common Mark Stud 37(3):407–428

Davey P (2006) The Nationals: the Progressive, Country and National party in New South Wales 1919 to 2006. The Federation Press, Sydney

Dowding K (1994) The compatibility of behaviouralism, rational choice and ‘new institutionalism’. J Theor Polit 6(1):105–117

Dowding K (1995) Model or metaphor? A critical review of the policy network approach. Political Stud XLIII(1):136–158

Dowding K (2001) There must be end to confusion: policy networks, intellectual fatigue, and the need for political science methods courses in British Universities. Political Stud 49:89–105

Dunsdorfs E (1956) The Australian wheat-growing industry 1788-1948. The University Press, Melbourne

Easton D (1953) The political system: an inquiry into the state of political science. Alfred A Knopf, New York

Flinn WL, Johnson DE (1974) Agrarianism among Wisconsin Farmers. Rural Sociol 39(2):187–204

Godfroij A (1995) Public policy networks: analysis and management. In: Kickert WJM, van Vught FA (eds) Public policy and administration sciences in the Netherlands. Prentice Hall/Harvester Wheatsheaf, London

Grant W, MacNamara A (1995) When policy communities intersect: the case of agriculture and banking. Political Stud XLIII(3):509–515

Griswold AW (1946) The agrarian democracy of Thomas Jefferson. Am Polit Sci Rev 40(4):657–681

Hall P, Taylor RCR (1998) The potential of historical institutionalism: a response to Hay and Wincott. Political Stud XLVI:958–962

Hay C (2006) Constructivist institutionalism. In: Rhodes RAW, Binder SA, Rockman BA (eds) The Oxford handbook of political institutions. Oxford University Press, Oxford

Hay C, Wincott D (1998) Structure, agency and historical institutionalism. Political Stud XLVI:951–957

Heathcote RL (1973) Drought perception. In: Lovett JV (ed) The environmental, economic and social significance of drought. Angus & Robertson, Sydney
Heclo H (1978) Issue networks and the executive establishment. In: King A (ed) The New American political system. American Enterprise Institute for Public Policy Research, Washington, DC
IC (Australia. Industry Commission) (1994) The tobacco growing and manufacturing industries. Report no 30 Australian Government Publishing Service, Canberra. http://www.pc.gov.au/__data/assets/pdf_file/0003/6969/39tobacc.pdf
Kickert W (1997) Public management in the United States and Europe. In: Kickert WJM (ed) Public management and administrative reform in Western Europe. Edward Elgar, Cheltenham
Lindblom CE (1959) The science of "muddling through". Public Adm Rev 19:79–88
Lindblom CE (1965) The intelligence of democracy: decision making through mutual adjustment. Free Press, New York
Mahoney J (2000) Path dependence in historical sociology. Theor Soc 29:507–548
Marsh D, Rhodes RAW (1992) Policy communities and issue networks: beyond typology. In: Marsh D, Rhodes RAW (eds) Policy networks in British government. Clarendon, Oxford
Marsh D, Smith MJ (2001) There is more than one way to do political science: on different ways to study policy networks. Political Stud 49:528–541
McAllister I (2009) Public opinion towards rural & regional Australia: results from the ANU Poll. Report 6, October 2009
McKay DH (1972) Stabilization in Australian agriculture. In: Throsby CD (ed) Agricultural policy: selected readings. Penguin Books, Ringwood
Meyer JA, Califano MG (2006) Good Intentions Corrupted: The Oil-for-Food Scandal and the Threat to the U.N. Public Affairs, New York
Mill JS (1893) Principles of political economy. D Appleton and Company, New York
Montmarquet JA (1989) The idea of agrarianism. University of Idaho Press, Moscow
Morgan D (1979) Merchants of grain. The Viking Press, New York
Morriss WE (1987) Chosen instrument: a history of the Canadian Wheat Board, the McIver years. Reidmore Books, Edmonton
Peters BG (1999) Institutional theory in political science: the 'new institutionalism'. Pinter, London
Peters BG, Pierre J (1998) Institutions and time: problems of conceptualization and explanation. J Publ Admin Res Theory 8(4):565–583
Peterson J, Bomberg E (1999) Decision-making in the European Union. St Martin's Press, New York
Pierson P (2000) The limits of design: explaining institutional origins and change. Governance 13(4):475–499
Pierson P (2004) Politics in time: history, institutions, and social analysis. Princeton University Press, Princeton
Richardson JJ, Jordan AG (1979) Governing under pressure: the policy process in a post-parliamentary democracy. Robertson, Oxford
Sabatier P (1988) An advocacy coalition framework of policy change and the role of policy-oriented learning therein. Policy Sci 21:129–168
Sanders E (2006) Historical institutionalism. In: Rhodes RAW, Binder SA, Rockman BA (eds) The oxford handbook of political institutions. Oxford University Press, Oxford
Sewell T (1992) The world grain trade. Woodhead-Faulkner, New York
Shepsle KA (1989) Studying institutions: some lessons from the rational choice approach. J Theor Polit 1(2):131–147
Shepsle KA (2006) Rational choice institutionalism. In: Rhodes RAW, Binder SA, Rockman BA (eds) The Oxford handbook of political institutions. Oxford University Press, Oxford
Smith MJ (1990) The politics of agricultural support in Britain: the development of the agricultural policy community. Dartmouth, Aldershot
Smith MJ (1991) From policy community to issue network: *Salmonella* in eggs and the new politics of food. Public Admin 69:235–255
Stewart J (2006) Value conflict and policy change. Rev Policy Res 23(1):183–195

Streeck W, Thelen K (2005) Introduction: institutional change in advanced political economies. In: Streeck W, Thelen K (eds) Beyond continuity: institutional change in advanced political economies. Oxford University Press, Oxford

Studlar DT (2005) The political dynamics of tobacco control in Australian and New Zealand: explaining policy problems, instruments and patterns of adoption. Aust J Polit Sci 40(255–274)

Thacher D, Rein M (2004) Managing value conflict in public policy. Governance 17(4):457–486

Waterhouse R (2005) The vision splendid: a social and cultural history of rural Australia. Curtin University Books, Fremantle

Whitwell G, Sydenham D (1991) A shared harvest: the Australian wheat industry, 1939-1989. Macmillan Australia, South Melbourne

Woodward D, Curtin J (2009) Beyond *country to national*. In: Botterill LC, Cockfield G (eds) The National Party: prospects for the great survivors. Allen & Unwin, Sydney

Chapter 2
Australian Wheat Industry Policy in Context

Keywords Australia • Wheat industry history • Wheat production • Wheat exports • Rural policy history

This chapter sets out to provide some contextual information about the Australian wheat industry to set the scene for the story and analysis that follow. First it provides some factual information on the Australian wheat industry; its geographic location, the nature of the industry and its contribution to Australia's economy. The industry is a major contributor to Australia's export earnings and has also played an important role in the development of much of the country's cultural and political life. The second part of the chapter provides some historical rural policy context for the discussion of the evolution of the statutory marketing arrangements for wheat that emerged in the 1940s.

Australian agriculture in the early twenty-first century receives low levels of government assistance and there is very little government involvement in the production or marketing of agricultural produce by way of regulation. However, this has not always been the case. The trend towards deregulation of agriculture began in the 1960s and 1970s. Prior to that time, as in other developed economies, the Australian Government intervened extensively in the market for agricultural produce. Collectively, the policy instruments employed were known as 'stabilisation' or 'orderly marketing' and included price supports, differential home consumption prices, guaranteed minimum prices and government underwriting. Where possible, restrictions on substitutes were introduced. This was the case with margarine, the import of which was banned unless the product was dyed pink. As later chapters describe, the wheat industry was generally successful in resisting the push for deregulation and retained key features of the 1940s stabilisation approach until the end of the twentieth century.

L.C. Botterill, *Wheat Marketing in Transition: The Transformation of the Australian Wheat Board*, Environment & Policy 53, DOI 10.1007/978-94-007-2804-2_2,

Wheat in the Australian Economy

In terms of area under cultivation, volume of production and value, wheat is Australia's most significant crop (Australia. Productivity Commission 2010: 9). In 2009 Australia produced nearly 21 million metric tonnes of wheat, of which 14.7 million metric tonnes were exported (ABARE 2009: 214). Wheat is consistently one of the country's top three rural exports, along with meat and wool, and is therefore an important contributor to national wealth. Although Australia is not as dependent on rural exports today as it has been in the past, they have still accounted for between 10% and 20% of export income over the past decade (ABARE 2009: 6). Australia produces 3% of the world's wheat crop on average but contributes 12% to the world wheat trade (Australia. Productivity Commission 2010: 17). Distinctively by world standards, wheat production in Australia is characterised by highly specialised, large-scale, capital-intensive, owner-operated farming systems oriented mainly to its overseas markets. The Australian wheat industry is diverse, ranging from smaller farm businesses which combine wheat growing with other activities, generally sheep raising and/or in association with other winter cereals, to large export-oriented businesses. It is characterised above all by its extensive nature, its low production per hectare — on average its yields are barely a third or less of those achieved by most western European nations — and its high productivity per labourer. Its development had to confront and overcome constraints imposed by extremely variable weather patterns in a generally semi-arid climate, often fragile and infertile soils and long-term geographic isolation.

The 7.7 million km^2 Australian land mass is roughly the size of the continental United States and lies between latitudes 10 degrees 41 minutes (10° 41′) and 43° 38′ south and between longitudes 113° 09′ and 153° 38′ east. It is the lowest, flattest and the driest of the inhabited continents. Its geological history is extremely prolonged, and unlike most of the northern hemisphere land masses, is measured in many millions of years. Recent geological events are confined to occasional so-called intraplate earthquakes, disturbances that occur in the stable portions of continents, and to basaltic lava flows associated with mostly extinct volcanoes. Consequently the present Australian landforms result from largely uninterrupted processes of prolonged wind and water erosion that provide the physical basis for the distribution and nature of biological and human activity in Australia. Many Australian soils that have evolved on these landforms are ancient, strongly weathered and infertile by world standards though those on floodplains and those developed on the basaltic lava flows are younger and more fertile. Most have surface layers that contain low organic matter and are often poorly structured and lack the qualities for sustained agricultural production without significant management inputs.

Australia's size and latitudinal range lead to the presence of a variety of climates — monsoonal, savanna, humid temperate, Mediterranean, steppe and desert. In terms of the Köppen climate classification, one of the most widely used systems, approximately 47% of the continent lies within the arid (BW) zone and a further 32% in the semi-arid or steppe (BS) zone. The dominant climatic variable relevant to agriculture is undoubtedly that of rainfall. The extensive, arid continental core is surrounded by vast areas of semi-arid lands where rainfall has a strong seasonal component of winter dominance in the south and summer dominance in the north, but which is characterised by high annual and monthly variability. In her poem, *My Country*, Australian poet Dorothea Mackellar wrote of a land of 'droughts and flooding rains' (Mackellar undated) and this remains an accurate portrayal of the uncertainty with which Australian farmers live. Australia has one of the most variable rainfall regimes in the world and, particularly in the east, is affected by the El Niño Southern Oscillation phenomenon — a wide scale ocean–atmosphere system centred in the tropical Pacific Ocean which has near global impacts (Lindesay 2003: 26). Australia's indigenous people adapted to water scarcity, managing the resource with 'care and restraint' (Rose 2005: 37). As Rose (2005: 40) points out, 'Aboriginal people spaced themselves across the continent in densities that reflect the rainfall of a given area'. As a survival strategy this was an effective means of living in the Australian environment and adapting to its limitations.

Europeans arriving in the late eighteenth century brought with them their own conception of climate, based on the relative reliability of Europe where rainfall is seasonal, unlike Australia's rainfall patterns which are so dependent on ocean currents. The colonists set about establishing the style of agricultural activity they had brought with them from their places of origin. An early note of warning was sounded by John Bigge in 1823 when, observing the 'uncertain climate', he reported back to Britain that the future of the colony

> will be that of pasture rather than tillage, and the purchase of land will be made with a view to the maintenance of large flocks of fine-woolled sheep; the richer lands, which will generally be found on the banks of the rivers, being devoted to the production of corn, maize and vegetables. (Bigge 1966 [1823]: 92)

In spite of this advice, a strong agricultural industry developed in Australia with high productivity growth and considerable ingenuity as farmers adapted to the Australian conditions. Droughts of varying magnitude have troubled agricultural producers on a regular basis. Significant drought events include the 'federation drought' of 1895–1902 and droughts in 1914–1915, 1937–1945, 1965–1968, 1982–1983, 1991–1995 (Lindesay 2003: 40) and the most recent long drought which began in 2002. This latest drought was followed in 2010 by a very wet La Niña event which saw large scale flooding across eastern Australia. The Australian Government responds to drought within a risk management framework and based on the principle that drought is a normal part of the farmer's operating environment. Government drought assistance is available to agricultural producers but it is framed in terms of supporting long term viable farm businesses within a policy paradigm of facilitating ongoing structural adjustment. (For detailed discussion of Australia's drought policy see Botterill 2003; Botterill and Fisher 2003; Botterill and Wilhite 2005). As is the case with farmers across the developed world, Australian producers have faced declining farm terms of trade, requiring ongoing productivity improvements in order to stay viable. This has exacerbated the impact of climate variability. Heathcote (1994: 100) sums up this challenge in his observation that 'the same rainfall which gave a bonanza wheat crop ... in the 1880s, would [have been] classed as a drought in the 1980s'.

Wheat was first grown in Australia in Sydney in 1788, by convicts with little practical agricultural experience, on 9 acres (3.6 ha) at Farm Cove, on the site of the future Royal Botanic Gardens. The first grain was harvested in July 1788 but most of the crop failed and in November 1799 the attempt was replaced by a 40 acre (16 ha) government farm established at Parramatta, now a suburb of Sydney. By 1799, 6,000 acres (2,400 ha) in the colony were under wheat, grown by free settlers who had arrived in 1793 but were not particularly successful farmers. As Dunsdorfs (1956: 10) points out, their lack of success was likely attributable to the fact 'that they had no idea of the hardships of pioneering of which they had not been adequately forewarned in England'. The spread of the Australian wheat industry from its small beginnings in the vicinity of the initial European settlement is examined by Dunsdorfs (1956) and summarised for the 1860–1939 period by Whitwell and Sydenham (1991) and is therefore not covered here in detail. From the initial cultivation in New South Wales, wheat growing spread out with the expansion of settlement and its associated market growth and the development of transport facilities.

Following the early production in New South Wales, wheat growing was taken up in Tasmania which was the first of the colonies to achieve self-sufficiency in 1815 or 1819 (Henzell 2007: 8). According to statistics compiled by Dunsdorfs (1956: 532–533), it was not until the early 1830s that Western Australia was recording a wheat crop, followed by South Australia and Victoria later that decade. Queensland's first wheat acreage was recorded in 1860. The areas under wheat in South Australia and Victoria grew rapidly; by 1860 both were producing more than New South Wales and this pattern continued into the twentieth century. The growth in the area under wheat in Western Australia 'took off after the turn of the century' (Tull 1991: 4), expanding markedly during the First World War. During the course of the twentieth century Australian wheat production increased tenfold.

The development of wheat-growing into a major agricultural industry, with an increasing emphasis on exporting, rested heavily on mechanisation and the breeding of varieties adapted to low and erratic rainfall. Mechanical innovations that greatly facilitated soil preparation, cultivation and harvesting all made their appearance in Australia in the second half of the nineteenth century and were widely adopted in the first years of the twentieth century. The outcome of these developments, along with government investment in railways, larger properties and recognition of the importance of fallowing and fertilisers, was the increase in production and the accompanying growth in the incomes of farmers anxious to overcome the problems of scarce and expensive labour.

In general terms, wheat growing areas are determined by soil type and fertility, topography, and rainfall with an annual average of between 400 and 600 mm per year falling predominantly during the winter and spring months in Australia. Suitable conditions occur on the mainland in an arc (the 'wheat belt') that extends from central Queensland west of the Dividing Range, through New South Wales and Victoria, to South Australia and picks up again in the southwest of Western

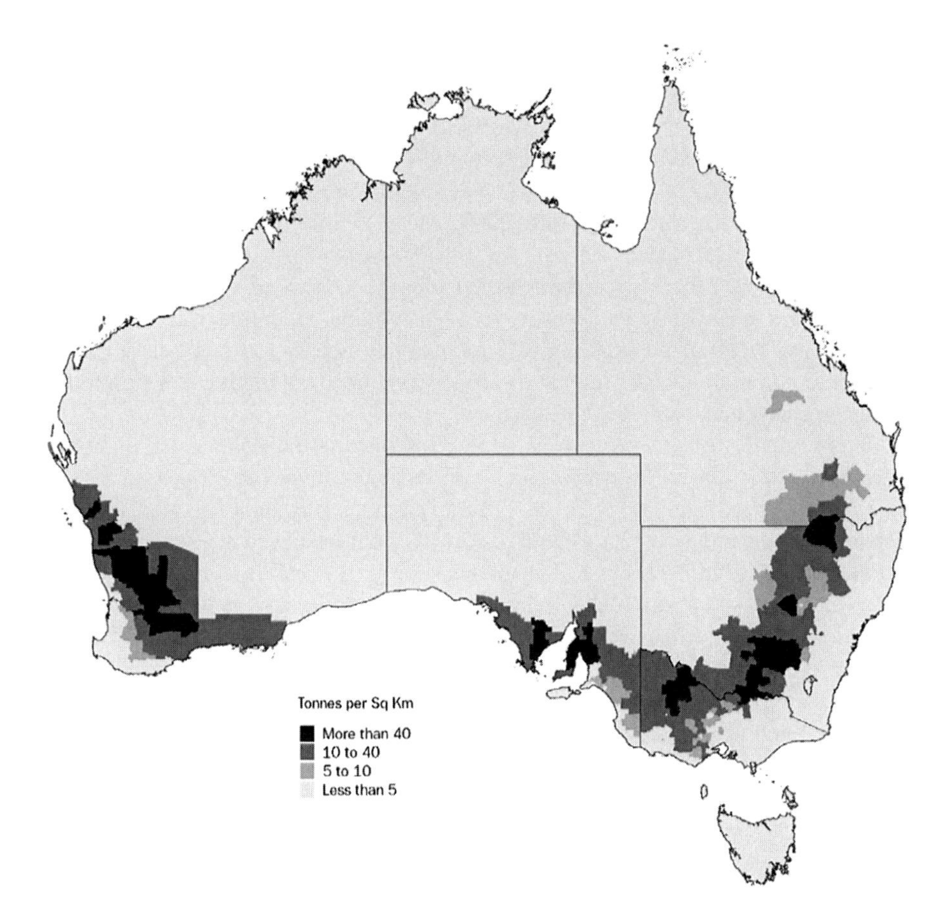

Fig. 2.1 The wheat belt, 2006: intensity of wheat production in Australia, tonnes produced per square kilometre (Source: Australia. Productivity Commission 2010: 54; copyright Commonwealth of Australia reproduced by permission)

Australia (Fig. 2.1). Wheat is grown also in a small area in Tasmania. Over the 39-year period from the crop year 1970–1971 until that of 2008–2009 there were substantial increases in both the area sown to wheat and in the volume of production, though with considerable variation from year to year according to local conditions (ABARE 2009). The 1970–1971 area of 6.48 million ha had more than doubled to the 13.2 million ha recorded in 2008–2009. Annual production was clearly related to the areas sown but also to yields which varied around a national average of 1.50 tonnes per hectare (t/ha) for the 39 years. During that period a maximum average yield of 2.11 t/ha was achieved in 2001–2002 and a drought-affected minimum of 0.77 t/ha in 1982–1983.

As might be expected, the volumes produced in the various states show strong correlations with the areas sown though differing fluctuations in year-to-year yields have produced changes in state production rankings. Throughout the period

1995–1996 to 2008–2009, for example, Western Australia has consistently recorded the largest area under wheat, approximately 35% of the total national area, followed in turn by New South Wales, South Australia, Victoria, Queensland and Tasmania. Volume of production has followed the same ranking except for the 1996–1997, 2000–2001 and 2001–2002 seasons when production in New South Wales exceeded that in Western Australia (ABARE 2009: 215). Yields followed a less regular pattern though; the average yield of 3.77 t/ha in Tasmania, with a cool temperate climate, was more than twice the national average of 1.69 t/ha over this period. Tasmania had the highest yield in every crop year without exception over the 14-year period cited. Victoria, also a state with a more dependable wet growing season with a maximum rainfall in late winter and early spring, recorded yields exceeding those of the other mainland states in most seasons.

The Western Australian wheat belt occupies a crescent shaped area with light but mixed soils that extends between Geraldton in the northwest to Hopetoun on the south coast. Productivity varies considerably across this large planted area of between 4 and 5 million hectares with maximum yields achieved in a strip north and south of the shire of Williams to the southeast of Perth. One of its major strengths has been attributed to its higher rainfall reliability and predictability of climatic conditions by comparison with the eastern states (DAFWA 2009: 1). However salinity is more serious in Western Australia than in other states with government, communities, research centres and farmers working together to manage the problem. The wheat belt of New South Wales extends in an arc from the Queensland border southwards as far as the Victorian border along the western slopes of the Great Dividing Range with its most productive region lying east and west of a line running from the Liverpool Plains to the Leeton Shire. Most of the western half of Victoria is included in the wheat belt which grows premium white wheat with productivity generally highest in the Mallee, the Wimmera and between Bendigo and Albury. The Victorian wheat belt merges into the South Australian wheat area, which extends east and west of Spencer Gulf. Most production is located in the Eyre Peninsula, the Yorke Peninsula and north and east of Adelaide as far as the lower Murray. Wheat production in Queensland is largely concentrated on the Darling Downs north of the border with New South Wales. Central Queensland contains smaller areas of more localised production.

Wheat Exports

By the early 1870s Australia had moved from being a net importer of wheat to becoming a net exporter and in 1873, Great Britain became the primary destination for wheat exports (Dunsdorfs 1956: 167). Although Australian wheat exports had been relatively unimportant in the nineteenth century, they kept pace with the industry's rapidly increasing production in the twentieth. However, Australia's position as a leading exporter was not so much due to the size of the industry, which remained modest by world standards, but because of the relatively small national consumption (Henzell 2007: 32–33).

With the exception of the Second World War and the immediate post-war years of the 1940s, when India and New Zealand replaced the UK as the principal destinations for Australian wheat, the UK was the dominant market, a situation that was particularly strengthened in 1932 by the Ottawa Agreements. These provided for quotas of meat, wheat, dairy goods and fruit from the Dominions to enter Britain free of duty in an effort to help counter the effects of the Great Depression. After the Second World War, although the UK resumed its predominance throughout the 1950s, the benefits were steadily eroded. With the prospect of British entry into the European Economic Community (EEC – later the European Union), the agreements became increasingly dispensable until the guaranteed UK market was finally lost in 1967 when the country was accepted as a member of the EEC. In no crop year since 1971–1972 has the UK appeared amongst the major destinations as it has sourced its wheat imports since then from other members of the European Union and Canada.

The post-war period saw a rapid increase in the global demand for wheat by an increasingly wide range of countries. General population growth, the world-wide increasing rate of urbanisation, relative prices, the availability and growth of domestic wheat supplies, changing tastes and a range of other factors have influenced the rate of increase in the demand for wheat (Whitwell and Sydenham 1991: 231). Following the loss of preferential status for wheat exports in the UK market, Australian researchers sought to create additional wheat varieties especially to meet the end-use requirements of Middle Eastern and Asian customers.

The international wheat market has become highly segmented with wheat requirements varying from country to country and even within individual countries. Australian wheat enjoys an excellent reputation for quality in international markets and the hard white varieties are particularly suited to the production of food products in East Asia, such as instant and fresh noodles. Wheat-based products have also increased in popularity in several countries where traditionally there has been a heavy reliance on rice as the staple food. Japanese consumption of bread, for example, increased rapidly in the 1950s and 1960s (Whitwell and Sydenham 1991: 256) and Japan became the second or third largest destination for Australian wheat in 13 of the 20 crop years in the 1960s and 1970s.

For the whole of the 1960s China was the leading market by far for Australian wheat exports after which the then USSR and Egypt shared its predominant position. However, of particular significance to the story in this book is the appearance of the Middle East as a significant new market for Australian wheat. Iran had appeared as the fifth largest destination in the 1964–1965 crop year and was the second largest in 1986–1987 and 1987–1988 (Whitwell and Sydenham 1991: 232). Iraq first appeared among the top five destinations in the 1970s. By the first decade of the twenty-first century, Iran and Iraq were Australia's leading markets for wheat and flour and, in 2001–2002, together absorbed over 28% of its total exports. Indonesia and Japan were also in the market as major purchasers, taking 12.7% and 7.3% respectively of the exports in that year.

Australian wheat is now shipped to more than 40 countries mostly in Asia and the Middle East. In the 2008–2009 season, the largest destination was Indonesia which took 20.3% of the total exports, followed by Iran which took 11.8% following a number of seasons (presumably when sanctions were in force) in which its purchases ceased or were very small. Other significant purchasers were Malaysia, Japan, the Republic of Korea and Yemen. China ceased to be a major customer in most seasons over the past decade and took less than 2% of Australia's wheat exports in 2008–2009.

The National Contribution

The precise extent of the contribution of the Australian wheat industry either to the national economy or to employment can be cited only in very general terms, given both the season-by-season variability and price fluctuations created by market conditions and exchange rates. Further, the availability of particular data may not always be consistent in the time periods to which they refer and are often up-dated. It should be noted, therefore, that data cited in this section are the best available at the time of writing and should be read with these qualifications in mind.

In 2009 Australia's national economy was valued at $US920 billion at the then current exchange rate (World Bank 2011). The Australian economy is dominated by its services sector which was estimated as accounting for 68.4% GDP in 2008, with industry contributing 29% and agriculture 2.5% (World Bank 2011). In 2007–2008, 125,594 Australian farms were solely dedicated to agricultural production, and 2007–2008 gross value of Australian farm production (at farm gate) totalled $A43.6 billion of which wheat contributed $A5.2 billion or approximately 0.5% of GDP. The high productivity of Australia's farmers has been a major factor in the success of the diverse wheat industry with its range from smaller farm operators to large export-oriented businesses. The wheat-sheep operations vary their production mixes in response to price signals and climatic conditions so the number of farms that would be included in wheat statistics varies from year to year. In 2008–2009, approximately 26,000 farm businesses grew wheat (Australia. Productivity Commission 2010: 52), however, this number disguises the highly concentrated nature of the production. In 2005–2006, 50% of wheat growers accounted for less than 10% of the crop while the top 10% produced almost half of Australian production (Australia. Productivity Commission 2010: 9).

The divide between the large and small producers roughly coincides with the split between those growers who produce for the export market and those who are focused on the domestic market. The largest wheat farms are in Western Australia and 90% of that state's crop is exported (Australia. Productivity Commission 2010: 9). Inevitably, the interests of these two groups, the large and small producer, diverge, particularly on the subject of export marketing arrangements. As will be discussed in later chapters, for many years the voices of the smaller producers have dominated the policy debate and it is their interests that were best represented by the collective ethos of the marketing arrangements in the period 1948–2008.

The contribution of wheat growing to total employment is clearly much greater than that made merely by the farm businesses themselves. Australian agriculture has important linkages with other sectors of the economy and consequently contributes to these flow-on industries. Whitwell and Sydenham (1991: 1) draw attention to this multiplier effect by stating that wheat production 'is a shared harvest…in the sense that there are a great many people whose employment is derived either wholly or partially from the activities of the wheat farmer'. These activities are crucial to the survival of many country towns 'whose inhabitants derive their livelihood by servicing the requirements of the neighbouring wheat farmers and their families' (Whitwell and Sydenham 1991: 1).

The Rural Policy Paradigm

As noted above, Australian rural policy settings prior to the 1970s were highly interventionist and it was in this context that government policy towards the wheat industry was developed. This section provides a brief overview of the development of European-style agriculture in Australia and the prevailing paradigm that dominated agricultural policy thinking until the 1950s.

European-style agriculture in Australia had a relatively inauspicious start.[1] Initially, the production of food and fibre was aimed at sustaining the convict colonies and the early farmers had little or no farming experience. The first Europeans to engage in agriculture were soldiers and emancipists, or freed convicts, who were given plots of land. As early as October 1788, Governor Phillip was writing to England seeking free settlers to cultivate the land so the colony could support itself (Clark 1950: 63). Emancipated convicts were granted small tracts of land for farming by the British Government and alongside these small settlers was a large group of wealthy officers who quickly recognised the potential for grazing rather than farming. Although the graziers invested in wool in the early nineteenth century and began to build the foundations of what was to become for many years Australia's most successful export industry, their achievements in the early years should not be overstated; until 1830 whale and seal oil were more important exports from the colony than wool. From the 1820s there was a speculative boom in wool production, however agriculture was still struggling.

In 1831 the British Government stopped granting land and started selling it. However, the graziers ignored these regulations and simply 'squatted' on Crown land without title. In the 1860s land reforms were introduced to break the squatters' monopoly and to further the idea of an industrious yeomanry in Australia by the creation of family farms. The 1860s land reforms were not overly successful as the wealthy squatters were in a good position to buy the land that was for sale and

[1]Some of the material in this section was previously published by the author in the *Taiwanese Journal of Australian Studies* Vol VI, 2005 and is reproduced with the permission of the publisher.

between 1860 and 1890 large areas of New South Wales and Victoria were bought by squatters. After the First World War, returning soldiers were given grants of land to thank them for their war service. These blocks were small and often of poor quality. These soldier settlement schemes were repeated after the Second World War and sowed the seeds for many of the adjustment problems that were to follow for Australian agriculture in the late twentieth century.

The 1920s saw an acceleration in government intervention in agriculture. A high home price scheme was introduced for butter to offset low export prices, a similar arrangement was in place for sugar growers and dried fruit attracted a high home price and an export subsidy. Between 1932 and 1936 bounties were paid to wheatgrowers and in 1938 a two price scheme was introduced for wheat. Shaw has observed that 'except for the pastoralist … practically every form of farming activity was receiving direct government help of some kind during the 1930s' (Shaw 1967: 17).

In 1943, a Rural Reconstruction Commission was established to inquire into

(a) The organization of Australian rural economy for the purposes of the defence of the Commonwealth and the effectual prosecution of the war, including the efficiency of methods of production, distribution and marketing of primary products, and the conservation, maintenance and development of the natural resources of Australia; and
(b) The re-organization and rehabilitation of the Australian rural economy during the post-war period. (Rural Reconstruction Commission 1944: 2)

The Commission's First Report contained a bleak assessment of the state of Australian agriculture: 'Perhaps an apt picture of many farmers in 1943 is that of very tired men worried by 10 years of difficulty, perplexed by doubts as to their future, but grimly carrying on with the task of the day' (Rural Reconstruction Commission 1944: 31). The report was concerned not just with the economics of rural industries but also with the social conditions of farmers and their families. In its Fourth Report, the Commission singled out the wheat industry for particular attention, with a focus on 'marginal wheat areas'; land which had been allocated for wheat growing but which for various reasons proved unsuitable for that purpose.

The Commonwealth Government responded to the Commission report with a statement in 1946 on 'A Rural Policy for Post-War Australia'. Prime Minister Ben Chifley outlined the following objectives of Australia's rural policy:

(i) To raise and make more secure the levels of living enjoyed by those engaged in and dependent upon the primary industries.
(ii) To secure a volume of production adequate to meet domestic food requirements, to provide the raw materials for our developing secondary industries, and to enable an expanding volume of exports to pay for necessary imports.
(iii) To encourage efficient production at prices which are fair to the consumer and which provide an adequate return to the producer.
(iv) To develop and use our own primary resources of water, soil, pastures and forests in a way which conserves them and avoids damaging exploitation. (Chifley 1946: 2)

These objectives are notable for their combined concern with fair incomes for producers and fair prices for consumers. There is also a striking similarity between these objectives and those set out in Article 39 of the Treaty of Rome which forms the basis of the European Common Agricultural Policy. The main difference is principle (iv) of the Australian statement which, somewhat ahead of its time, is concerned with the sustainability of agricultural production.

In the early 1950s, Australian agriculture received a boost as demand for wool for the Korean War saw significant price rises. This, along with world food shortages, resulted in considerable prosperity for most of Australian agriculture. The buoyant Australian economy also resulted in an increase in import demand and the Government's response to the pressure on the balance of payments was to introduce policies to increase exports. This meant increasing farm output. Accordingly, in 1952, State and Commonwealth Ministers with responsibility for agriculture shifted the emphasis of agricultural policy from income stability to the expansion of exports in order to earn foreign exchange income to pay the growing import bill (McEwen 1952). Ministers announced a set of production aims with a target date of 1957–1958. Due to the good conditions prevailing in agriculture in the early 1950s these targets were in fact met 2 years early.

Although the objective of policy shifted in the 1950s from income support to the expansion of production, the basic structure of the stabilisation schemes remained, and policy was concerned with providing income security for farmers and stable prices for consumers. Governments were also concerned to protect agriculture from unfair competition. Throsby has described Australian agricultural support at this time as 'a bewildering array of policy instruments which directly or indirectly affect[ed] farm prices' (Throsby 1972: 13). The resulting collection of support measures was more diverse than either the European Common Agricultural Policy or US farm policy (Lloyd 1982: 364).

It was against this background of concern for farm incomes and government willingness to intervene in the markets for agricultural production that the debate over wheat marketing took place. The next chapter begins the story of wheat industry policy which is the focus of this book.

References

ABARE (Australian Bureau of Agricultural and Resource Economics) (2009) Commonwealth of Australia, Canberra

Australia. Productivity Commission (2010) Wheat export marketing arrangements. Productivity Commission inquiry Report, No. 51, Canberra, 1 July 2010

Bigge JT (1966 [1823]) Report on agriculture and trade in NSW. Libraries Board of South Australia, Adelaide

Botterill LC (2003) Uncertain climate: the recent history of drought policy in Australia. Aust J Polit Hist 49(1):61–74

Botterill LC (2005) Australian agricultural policy and the push for trade liberalization. Taiwan J Aust Stud VI:85–110

Botterill LC, Fisher M (eds) (2003) Beyond drought: people, policy and perspectives. CSIRO Publishing, Melbourne

Botterill LC, Wilhite DA (eds) (2005) From disaster response to risk management: Australia's national drought policy. Springer, Dordrecht, The Netherlands

Chifley JB, the Rt Hon (1946) A rural policy for post-war Australia. A statement of current policy in relation to Australia's primary industries, Bureau of Agricultural Economics, Canberra
Clark CMH (ed) (1950) Select documents in Australian history 1788–1850. Angus and Robertson, Sydney
DAFWA (Department of Agriculture and Food Western Australia) (2009) Overview of the Western Australian wheat flour industry and potential export opportunities. Bulletin 4797 State of Western Australia
Dunsdorfs E (1956) The Australian wheat-growing industry 1788–1948. The University Press, Melbourne
Heathcote RL (1994) Australia. In: Glantz MH (ed) Drought follows the plough. Cambridge University Press, Cambridge
Henzell T (2007) Australian agriculture. Its history and challenges. CSIRO Publishing, Collingwood
Lindesay JA (2003) Climate and drought in Australia. In: Botterill LC, Fisher M (eds) Beyond drought: people, policy and perspectives. CSIRO Publishing, Melbourne
Lloyd AG (1982) Agricultural price policy. In: Williams DB (ed) Agriculture in the Australian economy, 2nd edn. Sydney University Press, Sydney
Mackellar D (undated) My country. Official Dorothea Mackellar website: poetry archive. http://www.dorotheamackellar.com.au/archive.asp
McEwen J, the Hon (1952) Agricultural production aims and policy. Commonwealth Government Printer, Canberra
Rose DB (2005) Indigenous water philosophy in an uncertain land. In: Botterill LC, Wilhite DA (eds) From disaster response to risk management: Australia's National Drought Policy. Springer, Dordrecht, The Netherlands
Rural Reconstruction Commission (1944) A general rural survey. First report. Commonwealth Government Printer, Canberra
Shaw AGL (1967) History and development of Australian agriculture. In: Williams DB (ed) Agriculture in the Australian economy. Sydney University Press, Sydney
Throsby CD (1972) Background to agricultural policy. In: Throsby CD (ed) Agricultural policy: selected readings. Penguin Books, Ringwood
Tull M (1991) Shipping, ports and the marketing of Australia's wheat, 1900–1970. Working Paper No 53 Economics Program, Murdoch University, Perth
Whitwell G, Sydenham D (1991) A shared harvest: the Australian wheat industry, 1939–1989. Macmillan Australia, South Melbourne
World Bank (2011) Data by country: Australia. http://data.worldbank.org/country/australia. Accessed 24 June 2011

Chapter 3
The Birth of Collective Wheat Marketing

Keywords Collective wheat marketing • Origins of the Australian Wheat Board • International wheat market • Canadian Wheat Board • Rural policy making

The previous chapter provided an overview of Australia's rural policy settings until the late 1950s, indicating that there was a high level of government intervention in the markets for agricultural products with the dual goals of income stabilisation for farmers and fair prices for consumers. The wheat industry was particularly subject to intervention. Early schemes were put in place during both world wars followed by the establishment of the Australian Wheat Board on an ongoing basis in 1948. This chapter discusses the early debates over intervention in the wheat industry, culminating in the passage of the *Wheat Stabilization Act 1948* which set up the Australian Wheat Board, and is broadly structured around the themes of the book. It begins with a discussion of the institutional aspects of the wheat marketing arrangements and then considers the nature of the policy community concerned with wheat marketing. It identifies the values that dominated the policy debate at the time collective wheat marketing became institutionalised, followed by a brief discussion of the development of a close cousin of the Australian Wheat Board, the Canadian Wheat Board. The chapter then places these policy developments in the context of international developments in the area of wheat marketing and attempts by both wheat importers and exporters collectively to manage the international wheat trade to their mutual advantage.

A number of writers has argued that one of the limitations of the historical institutionalist approach lies in identifying the starting point for the analysis and thus the potential for infinite historical regress in trying to track the origins of a particular institutional trajectory. Peters (1999: 67) argues that 'the choice of the relevant date from which to count future developments will be crucial for making the case that those initial patterns will persist and shape policies in the policy area'. In this vein, Mahoney (2000: 511) suggests that the point of origin of path dependent analysis is a contingent historical event 'that cannot be explained on the basis of

L.C. Botterill, *Wheat Marketing in Transition: The Transformation of the Australian Wheat Board*, Environment & Policy 53, DOI 10.1007/978-94-007-2804-2_3,

prior events or 'initial conditions''. This issue is not as problematic as these writers suggest. If Thelen's (1999) points are accepted that institutions are 'the product of concrete temporal processes' (384) and the 'enduring legacies of political struggles' (388), it is not unreasonable to suggest that an institution is born at the point at which the struggle is resolved in favour of one or other of the value positions in the debate, although the origins of the institution could be traced back through the positions of the various protagonists. In the case outlined below, the Australian Wheat Board was formally created by the *Wheat Stabilization Act 1948*. There had been temporary wheat boards during both world wars, but the 1948 Act marked the end of the debate that had taken place during the 1920s and 1930s between free market-oriented grain traders and those graingrowers pushing for a collective marketing arrangement. The 1948 legislation represented a win for the supporters of collective marketing over the traders, and their values were embodied in the Australian Wheat Board.

The policy debates around the wheat industry reflected the fact that the agricultural policy community generally, and the wheat industry more specifically, was fractured. The policy communities/networks literature regularly cites agricultural policy making as representing an area dominated by a cohesive, tight 'policy community' with shared values and objectives, a shared understanding of the policy problem and a common conception of the scope of acceptable solutions (Daugbjerg 1999; Grant and MacNamara 1995; Peterson and Bomberg 1999: 138–141; Smith 1990, 1992). This type of agricultural policy community did not emerge in Australia until 1979 (Botterill 2005). Prior to this time, farmers were split over the role of government in agricultural production and often these splits were not only between industries but also among farmers in the same industry.

Marsh and Rhodes (1992: 251) suggest that policy networks are spread along a continuum from loose issue networks to closed policy communities. They describe a range of network types based on the nature of their membership, the level of integration between those members, the resources they hold and the distribution of power. The network of policy players around the wheat industry prior to 1948 is best understood as an issue network. It included a diversity of players from growers to grain traders and was divided over the appropriate form of wheat marketing for Australia. The industry later took on the characteristics of a closed policy community in which members shared basic values but this only occurred after the value conflicts were resolved in favour of the collectivists and embodied in the Australian Wheat Board. The values of agrarian collectivism won out over ideas of free market liberalism and concerns over the introduction of a legislated monopoly.

Institutional Development and Change in the Wheat Industry

Wheat growing had become an established industry in Australia by about 1860 (Royal Commission on the Wheat Flour and Bread Industries 1934: 10). Following the gold rush, the industry expanded as miners moved out of the goldfields into agriculture. Although world wheat prices climbed in 1920, Australian farmers did not benefit as much as their North American counterparts and then between 1920 and 1922 there was a slump in prices which persisted. The price decline was due to a mismatch between supply and demand caused by 'the unrelieved pressure year after year of an enormous surplus which world markets failed to absorb' (Royal Commission on the Wheat Flour and Bread Industries 1935: 17).

The First World War had seen the introduction of compulsory 'pooling' of the wheat crop, administered by an Australian Wheat Board, but this ceased with the 1920–1921 season. Pooling was an arrangement under which 'wheat delivered in any season formed a 'pool' and growers received an aggregate pool price which was determined by averaging returns from all markets' (Watson and Parish 1982: 343–344) with deductions for marketing and other costs associated with the pool's operation. The main objective of pooling is to reduce short-term fluctuations in the prices farmers receive. Patton (1937: 230) argues that 'large-scale pool marketing of wheat' was a 'distinctive Canadian and Australian development' and describes the three key features of a pooling arrangement as follows:

> It represents centralized, in contrast to individual, selling by producers. It involves payments to growers on a deferred but equalized basis, in contrast to outright cash payment at the time of individual sale. It implies collective assumption by producers (possibly shared by government) of the inevitable risks of market price fluctuations. (1937: 219)

At the time of the establishment of the war time pool, the Australian Prime Minister had made a commitment to merchants that there would be a return to the open market for wheat, which he honoured (Whitwell and Sydenham 1991: 42). After the war, farmers' organisations in the four wheat exporting states set up their own pooling arrangements which continued into the 1930s in Victoria, South Australia and Western Australia. Of these arrangements, the Western Australian voluntary pool was the most successful, to the extent that the Primary Producers' Association in that state, having supported compulsory pooling during the war, 'later revisited similar proposals as attacks on the co-operative movement' (Smith 1969: 118). This opposition to compulsory pooling from parts of the wheat growing community in Western Australia was to play itself out in the Senate when the Commonwealth Government attempted to introduce compulsory pooling in 1930. It was also reflected in debates over wheat industry structures some 70 years later.

By the second half of the 1920s, groups of wheatgrowers in several states were beginning to organise and promote the virtues of a home consumption price for wheat and a compulsory national pool, but there was no effective voice for the wheat industry as a whole. A series of ballots of growers in the states either failed to pass or to gain the required majority to implement compulsory pooling. By 1930, the wheat industry in Australia was in dire straits. The Commonwealth Government called a wheat conference in February 1930, coming away with the strong impression that it had 'the firm support of a majority of growers' for its plan (Smith 1969: 149), which it attempted to enact through the *Wheat Marketing Bill 1930*. The legislation proposed a guaranteed price for wheat and a system of compulsory pooling. The guaranteed price was seen as an important component of a

'Grow More Wheat' campaign which the government had initiated in an attempt to boost Australia's export income. Smith argues that while the guaranteed price was justified, the government failed to make a strong case for the compulsory pooling element of its proposal; he suggests that this part of the legislation was in fact ideologically driven:

> The government added compulsory pooling to the guarantee scheme because of Labor[1] policy. During the 1920s Labor had initiated or supported several pooling proposals and had encouraged groups among the farmers agitating for market reform. Thoughts of putting the wheat merchants out of business fitted well with Labor rhetoric and could be used to appeal to farmers' susceptibilities about grasping middlemen and vested interests. (Smith 1974: 51)

The government's attempts to build a coalition of support for the legislation among State governments and farmers failed and the debate in and outside the Parliament over the legislation was heated; Smith (1974: 52) describes it as 'vicious'. The Parliamentary debate was also lengthy and 'so complicated that even a statement of the barest outlines makes a long and tragic story' (Dunsdorfs 1956: 269).

Not only did the government find the issue difficult, the non-Labor parties split over the issue, both between the then United Australia Party and the Country Party and within the Country Party itself. The leader of the Opposition, John Latham was highly critical of the Bill, arguing that it was part of Labor's plan to socialise industry. He was also scathing of the war-time pooling arrangements, arguing that, although he had believed that in war time pooling arrangements were necessary, the system had suffered 'gargantuan losses, most astounding frauds, and lamentable corruption' (Latham 1930: 1428). Labor members responded to the criticism of the Bill by accusing the Opposition of doing the bidding of the merchants. The member for Angas, Joel Gabb, referred to a concerted campaign by merchants to bring down the Bill which included 'inspired articles … being published in the press, and farmers, evidently in the pay of the merchants, … going around the country organizing wheat freedom leagues' (Gabb 1930: 1827). The merchants were certainly opposed to the Bill, and were working through their agents to 'agitate among growers' in opposition to the legislation (Smith 1974: 54). The merchants' opposition was based on the linking of a guaranteed minimum price with a compulsory pool; they supported the former but were obviously opposed to pooling as it would lead to the end of their business. The Bill was ultimately defeated by a narrow margin in the Senate when two Country Party members from Western Australia voted against the scheme, reflecting the suspicion in that state of a plan 'proposed by a government not only Commonwealth but Labor' (Smith 1969: 121). According to Mitchell (1969: 13–14), the Minister for Markets in the Scullin Government was 'profoundly disappointed at the result after months of hard work' as 'The passage of a Wheat Marketing Act had been his life's ambition'.

[1]The spelling of the Australian Labor Party has varied over its lifetime. Although the party itself adopted the spelling "Labor" in 1912 (Australian Labor Party 2009), it continued to be spelt with the 'u' in much academic writing in the early half of the twentieth century and in Hansard, the official record of Parliamentary debates. For consistency, this book uses the modern spelling, except in the case of direct quotations where an alternative spelling is used.

The price of wheat continued to drop and wheatgrowers, denied the guaranteed price, became further indebted. In July 1931, the government again attempted to introduce wheat marketing legislation, this time including a compulsory pool but no guaranteed price. This legislation was also defeated in the Senate. According to Dunsdorfs (1956: 273), a result of the 1930–1931 policy debacle was that 'bankruptcy threatened nearly every wheat farmer'. The Labor government was defeated at a general election in 1931 and in 1934 the United Australia Party government, which was by then in coalition with the Country Party, set up a Royal Commission on the Wheat, Flour and Bread Industries. While it awaited the Commission's findings, the government passed two linked pieces of legislation to provide financial assistance to the wheat industry, the *Wheat Bounty Act 1934* and the *Wheat Growers Relief Act (No 2) 1934*.

The Royal Commission described the state of the wheat industry in 1935 as 'congested' (1935: 30) and stated that 'The world wheat position makes it difficult for the Commission to take an optimistic view of the prospects of the industry during the next few years' (1935: 32). At the time of the Royal Commission, farmers had three options for disposing of their harvest: selling to merchants at the day's cash price, storing it with a merchant or miller for later sale or selling it through one of the farmer-operated pools. These pools were receiving declining support by the time of the Royal Commission and, as Whitwell and Sydenham (1991: 36) put it, 'the pools were languishing and the merchant had re-emerged supreme'. Wheatgrower representatives continued to agitate for the introduction of a home consumption price and compulsory pooling and, in response to strong representations both for and against such an arrangement, the Commission gave detailed consideration in its Second Report to the issue of the reinstatement of compulsory pooling.

The Commission (1935: 171–177) outlined the advantages and disadvantages of a system of centralised control of wheat marketing as follows. It identified weaknesses in the existing system of farmers delivering their crops to merchants, particularly with respect to the masking of market signals which was seen to be contributing to the instability of world wheat markets. The Commission went so far as to argue that in 1931 'Australia was to some extent responsible for depressing the world wheat position as far as price was concerned' (1935: 172). It suggested that a centralised marketing arrangement would 'not have fallen into this error'. The second advantage put forward by the Commission related to the enforcement of international wheat agreements. Thirdly, the Commission suggested that economies of scale would result from a centralised marketing system which would avoid duplication of effort at sidings as well as savings on ship chartering. Fourthly, the Commission tied a centralised marketing system to the introduction of a home consumption price for wheat for human food, which it had recommended in its First Report. The Commission noted that farmers argued that the industry 'had a definite right to a home consumption price' (1935: 175) and the Commission agreed that, while other industries were subject to such schemes, it was in favour of a home consumption price for wheat as well. Fifthly, and the Commission gave particular emphasis to this point elsewhere in its report, the maintenance and improvement of the standards of Australian wheat was something that a centralised marketing system would achieve, including by making 'suitable rewards to such farmers as were producing grades and types of wheat which were above the average' (1935: 176). Finally the Commission pointed to greater efficiencies to be achieved in rail transport from a centralised system. The Commission also noted that a

> centralized organization would afford an opportunity for wheat-farmers to formulate a common policy on matters concerning their industry. At present there is no such common meeting ground, for, with due deference to existing farmers' organizations, there is today no widespread unanimity between them. (1935: 176)

The Commission also considered the potential disadvantages and 'dangers' of a compulsory centralised marketing system for wheat. First, it raised the concern that the home consumption scheme must not result in excessively high prices in the domestic market and argued that safeguards would need to be introduced to regulate such prices. It related this concern to its second issue – the potential for poor management of any new body. It was blunt in its assessment:

> The Commission is well aware of the fact that many organizations which are wholly controlled by farmers are frequently unsatisfactory. It is essential that the wheat consumer and the general public be represented on any body which is in any way concerned with the regulation of the local price of a commodity which is so vital to the human needs as bread. (1935: 177)

Thirdly, the Commission presciently pointed to the difference between the wheat industries in the various States; an issue which was to arise in debates in the 1990s over the restructuring of wheat marketing. The Commission highlighted the domestic focus of the eastern wheat-producing states compared with the export focus of growers in Western Australia and raised the issue of equity of treatment by a single organisation of the needs of these two groups. Fourthly, the Commission pointed out that a centralised marketing system would result in the demise of the existing private sector bodies and co-operative arrangements which had been built up to acquire and sell the Australian wheat crop. Fifthly, the Commission pointed out that the removal of competition would deprive farmers of the capacity to take their business elsewhere if they were dissatisfied with the service they were receiving. The removal of this 'safety valve' for grievances could, in the Commission's view, lead to an accumulation of discontent, resulting in 'very serious opposition which might easily result in political agitation culminating in the disruption of the system' (1935: 177). Finally, the Commission was concerned that applying the home consumption price to the whole crop in the non-exporting states of Queensland and Tasmania would result in an increase in otherwise uneconomic wheat growing in those states by artificially supporting the prices received.

On balance, the Commission recommended in favour of the establishment of a centralised marketing system for acquiring and selling wheat in Australia. It recommended that a poll of farmers be arranged to determine the level of support for such a scheme and that, if it were supported, the Commonwealth and State Governments move to set up a Commonwealth Wheat Marketing Board and State Wheat Marketing Boards. Whitwell and Sydenham (1991: 55) note that 'the coalition [government] rejected most of the commission's recommendations … before coming up with its own proposal'. When a scheme similar to that being considered by the government was ruled unconstitutional by Australia's High Court, the government's plan had to be abandoned. A referendum to overcome the constitutional obstacles failed and as a result of this and a rise in wheat prices in 1937, 'the federal government lost interest in formulating an alternative plan' (Whitwell and Sydenham 1991: 55). A home consumption scheme was finally achieved, without the compulsory pooling that the Royal Commission had felt was necessary, with the passage of the *Wheat Industry Assistance Act 1938*. The scheme involved both Commonwealth and State legislation which meant that both levels of government shared 'some of the political odium caused by the artificial inflation of domestic [flour] prices' (Whitwell and Sydenham 1991: 56). Further schemes were under consideration at the beginning of the Second World War.

According to Dunsdorfs' colourful history of wheat-growing in Australia, the 'general failure of Australian wheat-growers was averted only by the still greater calamity of the Second World War' (1956: 263). Under national security legislation, the Commonwealth set up an Australian Wheat Board responsible for marketing, storage and shipping and for compulsory pooling of wheat. Unlike the First World War arrangement, this wheat board included representatives of wheat growers, merchants and bulk handling authorities. However, changes to the Board's composition were made in 1940 and 1941 so that by the end of the war, the board was grower-dominated with no representation of the wheat merchants (Whitwell and Sydenham 1991: 60).

It is worth noting that the change of government in 1931 had not resulted in the Country Party's achieving greater success than Labor in arriving at stable and broadly accepted wheat policy settings. Part of the difficulty arose because of its coalition arrangement with the United Australia Party (UAP). The latter continued the opposition to compulsory pooling that it had demonstrated during the 1930 debate, considering it to be a form of 'socialism'. The UAP was also opposed to a tax on flour which was the mechanism under consideration for funding a wheat stabilisation scheme (Smith 1969: 268).

The Nature of the Policy Community

The wheat industry has played an important part in the evolution of Australia's political parties. Although generally regarded as a stable two-party system (see for example Sartori 1990: 340), Australia actually has three long standing political parties, the second oldest of which is the National Party of Australia, formerly the Country Party. Country Parties arose in the Australian states in the second decade of the twentieth century and the first national Country Party was established in the Commonwealth Parliament in 1921. These Country Parties were predominantly established in response to the grievances of wheatgrowers who regarded the two city-based parties, the Nationalists and Labor with distrust. However, the Country

Parties spent much of their time in coalition with the other non-Labor Party. This strategy of coalition was not without its opponents; for example in 1922 a split occurred in the state of New South Wales between the coalitionists in the Country Party and the 'True Blue' Progressives (Davey 2006: 29).

It was largely dissatisfaction with the performance of the Country Parties in coalition that motivated wheatgrowers to establish non-partisan organisations to further their interests. These attempts were mixed in their success. Not only did the federal structure fragment the growers but, within states, disagreements resulted in more than one grower organisation being formed. As a result the industry was fractured, with disagreement among and within wheatgrower organisations over the key policy issues such as compulsory pooling. The grower organisations also varied in their level of policy sophistication, with some such as the Wheatgrowers' Union of New South Wales remaining 'an ill-disciplined protest organisation' (Smith 1969: 224). Smith (1969: 220) reports that policy debates within the various bodies were frequently fraught; the debate over wheat policy at the 1939 conference of the New South Wales Farmers and Settlers' Association lasted 10 hours before a resolution was agreed. Even when conditions in the industry finally pushed the various organisations to adopt similar policies, they 'did not eliminate differences in style, vitality, and political skill' (Smith 1969: 255).

The government apparently mishandled the industry in the 1930–1931 debates around wheat marketing policy. This was not entirely its fault. It had attempted a consultation process with industry by holding conferences of wheatgrowers and State government representatives in 1930. However, it was faced with a 'divided constituency with no systematic capacity for giving expression to the views held within it' (Smith 1974: 56). Smith (1974: 64) argues that

> Its [the government's] bid for wheatgrower support left out the strength of wheat trading interests, the incapacity of wheatgrowers to express a united view, and the hostility of the Primary Producers' Association of Western Australia which was one of the strongest of existing growers' organisations.

The government had left the 1930 conferences believing it had the support of the majority of growers, however 'the proceedings of the conference had disguised the lack of communication between growers' organisations in the separate states and the unrepresentativeness of the new organisations' (Smith 1969: 149).

The industry soon recognised the desirability of seeking a single representative voice in the policy debate. Following the failure of the 1930 legislation, representatives of the New South Wales, Victoria, South Australia and Western Australia grower organisations met in February 1931 to discuss the formation of the Australian Wheatgrowers Federation (AWF). The meeting resolved that the objective of the organisation should be 'To unite growers into one Federal Organisation for the promotion and Defence of their industry'. Membership was open to 'State organisations of bona fide wheatgrowers, which are non-party political and non-trading' (Australian Wheatgrowers Federation 1931). The formation of the AWF did not result in its immediate recognition as the authoritative voice of industry on wheat policy. As late as 1938 and 1939, the Farmers and Settlers' Association of

New South Wales 'consistently took the initiative in dealing with governments, but did not work through the A.W.F.' (Smith 1969: 223). Similarly, the Western Australian Primary Producers' Association 'remained aloof from the wheatgrowers associations in the eastern states and made no attempt to join the A.W.F' (Smith 1969: 126–127). Following the defeat of the Scullin Labor government at the end of 1931, the incoming United Australia Party did little to engage with the Australian Wheatgrowers Federation on wheat policy, tending to consult with wheat traders and millers in the development of industry policy (Smith 1969: 171). Smith (1969: 182) reports on the ineffectiveness of the AWF in the following terms: 'The A.W.F. provided a forum for most organisations, passed copious numbers of resolutions, but failed to establish consistent connections with government'. Mitchell (1969: 33), however, takes a much rosier view, writing that, by the time of the AWF's 1938 conference, it had entered a 'new era ... for it became recognised by the Federal Government of the day as being the official mouthpiece of Australian wheatgrowers'.

The Birth of the Australian Wheat Board

The creation of the Australian Wheat Board was protracted, partly due to the organisational weakness of the wheat industry and partly because wheatgrowers were suspicious of collective action (Whitwell and Sydenham 1991: 30) both in terms of pressing their claims and with respect to the form of the government intervention they were seeking. Smith (1969: 44) highlights the ambivalence of growers to acting collectively, arguing that this type of action 'had no place in the growers' ideal world: when pressed into it many tended to revolt as soon as collective policies ran across their own short-term interests'. However, by 1939, there was a developing consensus that a collective approach in the form of compulsory pooling underpinned by a home consumption price was the best solution to the problems confronting the wheat industry. National security arrangements during the Second World War reacquainted the grains industry with orderly marketing but the development of the institutions from the 1939 Wheat Board to the introduction of the more stable arrangements in the *Wheat Stabilization Act 1948* proceeded only marginally more smoothly than the debates of the early 1930s. Mitchell (1969, Chapter 3) provides a detailed blow by blow account of the activities of the Victorian Wheat and Woolgrowers Association during the war years, reporting on increased levels of grower activity over the guaranteed price for wheat, including a 'monster meeting' at Bendigo in November 1939 and calls for direct action. One meeting went so far as to vote on a motion to

> accept the offer of the women of our wheatgrowing areas to cooperate with us in the endeavour to get justice for our industry and that they be invited to all public meetings on the wheat question, and be allowed to submit their own independent motions as they think necessary. (Mitchell 1969: 46)

Debate continued within the industry throughout the 1940s. In 1946, Prime Minister Chifley announced a stabilisation plan for the wheat industry but the Commonwealth recognised that any such scheme required cooperation by State governments (Mitchell 1969: 96). Gaining this cooperation took a couple of years. The industry was divided over the details of the Commonwealth proposal. Mitchell (1969: 97) reports that the October 1946 meeting of the Australian Wheatgrowers Federation 'saw a showdown on Wheat Federation policy in relation to the proposed wheat stabilisation'. In the end growers supported the plan which was not regarded as ideal but which 'they would rather accept … than … be once more at the mercy of the open market' (Mitchell 1969: 99). Although growers were generally supportive of the proposed scheme, their political counterparts in the Country Party were not as convinced of its merits. For example, in 1947 the Country Party in Victoria sought unsuccessfully to set up an alternative, state-based scheme (Mitchell 1969: 105).

The Australian Wheat Board, which was to last for six decades, finally came into being with the passage of the *Wheat Stabilization Act 1948*; although, as Dunsdorfs (1956: 294) notes 'its main principles had been in force since the end of the [second world] war'. He notes succinctly that these principles were '(a) Price Guarantee. (b) Home Consumption Price. (c) Stabilization fund. (d) Wheat Board' (1956: 296). In addition, the Act introduced an index of costs of production which was tied to the level at which the government guaranteed the price payable to wheat growers. The new Wheat Board was granted a monopoly on the marketing of wheat in both the domestic and export markets.

The Wheat Board was born in a particular political context and embodied the values of its time. It addressed concerns about the stability and levels of wheat grower incomes and removed that 'enemy' of the wheatgrower, the 'middle man', from the process of marketing wheat. Growers were suspicious that the wheat merchants were 'price destroyers' as they were not interested in getting the best price for the wheat they sold, just making the best profit (Harper 1928: 48). The free market was seen to be contributing to low prices because the Australian wheat crop was 'forced upon the market, regardless of demand' (Harper 1928: 48), thus robbing sellers of their bargaining power (1928: 49). This position was vindicated to some extent by the Royal Commission's suggestion noted above that the manner of the marketing of Australian wheat had depressed world prices. In spite of Opposition Leader Latham's claims in the Parliamentary debates of 1930 of mismanagement and fraud among the wartime pools, proponents of compulsory pooling argued that the pools had 'demonstrated beyond doubt that they are able to reduce the cost of handling between the farmer and the ship' (1928: 50).

By the 1940s, the collective principles embodied in compulsory pooling had won over the dominant individualism that had characterised the position of the more prosperous growers. By 1939 even the Primary Producers' Association of Western Australia, previously opposed to Commonwealth government action, had been converted to the idea of government intervention in the form of a home consumption process and a compulsory pool (Smith 1969: 239). By 1939 too the Australian Wheatgrowers Federation had become more effective at developing consistent policy positions and it had attained the recognition of government (Smith 1969: 296). Whitwell sums up the changed policy environment as follows:

> Several things stand out about the negotiations preceding the 1948 scheme. The first is that the wheatgrower organisations feature prominently. This was a major contrast with the negotiating process in the 1930s. The AWF was formed in 1931 with the aim of giving grower organisations an effective political voice but invariably it was ignored. Policy discussions in the 1930s largely took place at the ministerial level. The AWF had only a marginal impact. During World War II, however, wheat politics assumed a new significance and gradually, as wheatgrower organisations consolidated support, the AWF's importance increased. (Whitwell and Sydenham 1991: 63)

Dunsdorfs (1956: 291) attributes this greater role to the fact that the wheatgrowers had, in the inter-war period, 'acquired an incomparably better economic and policy status'.

The new Wheat Board arrangements reflected the general tenor of Australian rural policy at the time, for example as expressed in Chifley's 1946 *A Rural Policy for Post-War Australia* statement with its focus on fair prices for both consumers and producers and food security. The principle of grower control of the marketing arrangements had evolved over the period since the First World War; the second war time wheat board had not followed the pattern of the first which was comprised entirely of politicians. There were seven wheatgrowers appointed to the 1948 Wheat Board. Allied to the creation of the marketing monopoly was the regulation within the wheat growing states of the receival and bulk handling arrangements under state legislation. In the case of Queensland, a State Wheat Board was established and in other states the bulk handling responsibility was variously allocated to government agencies or co-operative bodies (Dunsdorfs 1956: 297). Whitwell (1993: 29) argues that the creation of the monopoly organisations in the states for the transport and handling of wheat was a 'quid pro quo' for the passage of state legislation which was required to underpin the price fixing component of the new stabilisation arrangements.

Institutions are the product of political compromises at a particular point in time and they reflect a particular set of values. This does not mean that these views are universally held, which is why, as Thelen and Steinmo (1992: 9) point out, battles over the establishment of institutions are often fought vigorously. Establishing institutions to give effect to a political compromise 'can save political actors the trouble of fighting the same battle over and over again'. After decades of debate over wheat marketing after the First World War, the 1948 Act provided predictability and allowed the grains industry to focus 'on extracting, by quiet tactical struggles over details and administration, maximum benefits from the scheme'(Smith 1969: vi). While the growers and the government seemed satisfied with the collective values and the level of government support incorporated in the 1948 legislation, others were more critical, particularly of the tendency to accept without question the proposition that the broader community had an obligation to alleviate the hardship facing wheat growers. Dunsdorfs (1956: 337) argues that 'The bounties and relief aid to wheat-growers during the depression, can be justified, because they were regarded as temporary measures to bridge over the depression and to save the industry from a collapse'. By contrast, Professor of Agriculture at the University of Melbourne, and member of the Rural Reconstruction Commission, Samuel Wadham (1944: 18) raised his concerns about the support of farmers in 1944 as follows:

> The financial support of uneconomic farming industries is merely a waste of people's time, patience and money. It arouses the hostility of the townsman and turns the farmer into a mendicant. When thinking of the history of wheat-growing in the past 10 years, during which the Australian public has given the wheat-farmers enough to hang on, one is forced to ask whether it would not have been better if we had, in 1930, made up our minds that we had made a mistake, cut our losses, bought the people off their properties and started them on something else that would have made them feel that they were doing something for the national good.

Similarly, Andrews (1936) was less inclined than Smith (1969: 8) to accept that conditions in the wheat industry were so bad that even the most successful needed assistance. He was prepared to contemplate the contribution to hardship of the 'personal inefficiency of the farmer' (1936: 112) and suggested an alternative to home consumption prices and 'paternalistic' government intervention. This would involve

> a reorganization of the industry so as to retain the efficient units and eliminate those which are hopeless or in which the risks are excessive; the efficient being those who can continue production under present conditions with any possible assistance in the lowering of costs which can be rendered by Governments without much sacrifice of their own revenue, and which can be arranged by cooperation between investors' and farmers' organizations and labour organizations. The remainder would be transferred to other rural activities or absorbed into other industrial branches. (Andrews 1936: 135)

This latter policy approach was clearly inconsistent with agrarian notions of the hard working farmer battling the elements with hardship generated through no fault of his own. As a basis for policy, concepts of structural adjustment such as this did not gain acceptance for a further half century or more.

A Parallel Case: The Establishment of the Canadian Wheat Board

Australia was not the only wheat producer which chose to adopt a policy of statutory marketing of its crop. Schmitz and McCalla (1979: 96) argue that 'The character of the Canadian Wheat Board is a unique function of Canadian history and institutions; as such, it should be judged only in that context'. However, the striking parallels between the development of the Australian Wheat Board and the creation of the Canadian Wheat Board suggest that a comparative perspective can be taken. The timelines of the two organisations in their early years ran almost in lock step and the rationales underpinning the development of pooling arrangements and export monopoly arrangements were very similar. The structure of the boards and the role of wheatgrower organisations have differed and, in more recent years, the institutional changes in Australia have been more profound. Nevertheless, there is value in briefly recounting the parallel history of the Canadian Wheat Board.

The defining events in the early history of both Wheat Boards were the two world wars and the depression. The other common feature of the two industries was their export dependence and exposure to international wheat prices (Whitwell and Sydenham 1991: 31). The Canadian wheat industry had developed roughly contemporaneously with the Australian industry, starting only a couple of decades later in the 1880s. The first Canadian Wheat Board was established in 1919 to manage the transition from war time monopoly marketing arrangements (Morriss 1987: 14); and according to Britnell and Fowke (1949: 629) was 'modelled on Australian experience'. The Board was dissolved in 1920 and growers, like their Australian counterparts, set up voluntary pooling arrangements. The conditions in the international wheat market which contributed to the withering of the Australian pools in the 1920s also impacted on the Canadian arrangement. By 1929, the Canadian pools were dependent on government support, first provincial then federal, to bail them out of financial difficulty (Schmitz and McCalla 1979: 81). The federal government introduced a stabilisation scheme in 1931 which in turn led directly to the Canadian Wheat Board legislation of 1935. The 1935 Board combined pooling with a guaranteed minimum price and was basically established to liquidate the surpluses that had accumulated in the provincial pools.

Although some analysts argue that growers were instrumental in the establishment of the Canadian Wheat Board (Britnell and Fowke 1949: 632; Fowke 1947: 241; Schmitz and McCalla 1979: 81), there were some growers who were less enthusiastic. The United Grain Growers Limited, a farmer cooperative involved in wheat marketing, recognised the need for emergency war time measures but never advocated a permanent monopoly Board (MacGibbon 1952: 192). MacGibbon suggests that the CWB was formed to address a series of problems confronting the grains industry; the Depression, severe drought in 1937 and the impact of war, including the loss of some export markets. He argues that 'National wheat policy has been shaped largely by attempts to solve the problems created by these circumstances one after another as they arose' (MacGibbon 1952: 221). This analysis is consistent with that of Morriss (1987: 14) who describes the development of the Board in terms which suggest it was a much more accidental and less politically fraught process than the evolution of the Australian Wheat Board:

> Born as an emergency expedient to meet an unparalleled combination of economic and environmental disasters in Western Canada, the board was at best an unwanted child in the eyes of its political father.

Public support was also equivocal and the Liberal government faced opposition from within its own ranks, as well as from the official opposition in the Parliament, to the establishment of a compulsory and permanent arrangement. MacGibbon (1952: 36) argues that 'The Wheat Board Act, when it finally passed Parliament, bore all the marks of a compromise measure'. This compromise saw the Wheat Board operating in parallel with the free market which resulted in what MacGibbon (p. 49) describes as 'an alternating system of marketing'; when prices were high, growers sold their wheat on the open market, when they dropped below the CWB's minimum delivery price, wheat was delivered to the Board. Any losses incurred by the Board were absorbed by the government. Britnell and Fowke (1949: 632) argue that the Board was established 'merely as an optional marketing channel which would free the producers from dependence upon the open market system without in any way interfering with that system'. In response to problems of overproduction in 1943, there was a complete change to this approach with the granting of a monopoly to the Board over wheat marketing (Britnell and Fowke 1949: 637). At the beginning of the twenty-first century, the Canadian Wheat Board retains its monopoly powers and its Act still refers to its object of 'marketing in an orderly manner, in interprovincial and export trade, grain grown in Canada' (*Canadian Wheat Board Act* S5). It has been under some pressure from the World Trade Organisation (WTO) (Furtan 2005: 98) and the *Revised Draft Modalities For Agriculture* under the Doha Development Round Agriculture Negotiations include reference to the removal of export monopolies from state trading enterprises by 2013, a provision which would cover the activities of the CWB (WTO 2008: 69).

The International Context

Australian and Canadian wheat growers were not alone in being spooked by the market conditions of the 1930s. The price of wheat had fallen 65% between 1929 and 1931 (Gordon-Ashworth 1984: 136). While exporting nations were concerned by the precipitous fall in prices, importers were concerned about security of supply. Economists similarly regarded the market conditions of the 1930s as justification for some level of intervention in the market for wheat. Writing in the early 1950s, Harbury (1954: 95) argued that embracing the free market would 'herald catastrophe' and that 'There is no need to elaborate time-worn arguments to support this'.

During the 1930s and 1940s a series of attempts was made to introduce an International Wheat Agreement with the apparently incompatible dual goals of improving income stability for producers and at the same time expanding world trade in wheat to ensure that supply was meeting demand (Harbury 1954; Tyszynski 1949: 27). In 1930 and 1931 alone 16 meetings were held to discuss international wheat market conditions (Gordon-Ashworth 1984: 136). Following several false starts, an International Wheat Agreement (IWA) finally came into force in 1949 to run for 4 years. It was renewed virtually unchanged in 1953 and 1956. Other agreements followed, leading the International Wheat Council to claim in 1991 that the 1949 agreement 'proved to be the first of a continuous series of agreements which have maintained a framework for international cooperation in wheat matters for over 40 years' (International Wheat Council 1991: 2). While this claim is true in that there has been a series of agreements, the distinctive features of the 1949 agreement that arguably led to its success were only characteristic of the first three agreements.

The 1949 Agreement was greeted at the time as a 'fresh approach' to international commodity agreements and 'a great experiment in world economic co-operation'

(Tyszynski 1949: 27). It was also considered to have 'none of the objectionable restrictive features that characterised pre-war commodity agreements (Hodan 1954: 225). Unlike the traditional approach to market management through the mechanism of buffer stocks, the IWA established a minimum and maximum price within which guarantees were made by both suppliers and importers. As Golay (1950: 443) explains:

> In substance, the International Wheat Agreement is a contractual obligation between 5 wheat exporting countries and 36 wheat importing countries, in which: (1) Each member exporting country, if called by the Wheat Council to do so, agrees to export a stated amount of wheat to member importing countries at the maximum price defined in the agreement. (2) Each member importing country, if called by the Wheat Council to do so, agrees to buy a stated amount of wheat from member exporting countries at the minimum prices defined in the agreement.

An important feature of the agreement was that it was not intended to cover all of the wheat trade. Trade which occurred outside the agreement – for example Argentina, while an important exporter, was not a signatory to the agreement – would be subject to a fluctuating price reflecting supply and demand. Harbury (1954: 82) argues that

> This dual price feature, together with the guaranteed quantities, gave the whole scheme something of the elegance of the theoretical economist's model and the attractiveness of a compromise between the twin aims of commodity policy – stability simultaneously with the adjustment of supply to changes in demand.

International commodity agreements require action by national governments to ensure their implementation (Gordon-Ashworth 1984: 93). With respect to exporting countries, Tyszynski (1949: 35) suggests four possible mechanisms: '(a) state trading, (b) subsidies to farmers, (c) export subsidies to traders, (d) internal buffer stocks'. The Australian Wheat Board's structure and mission was therefore highly compatible with the prevailing international approach to wheat policy; although Tyszynski does not see a monopoly as a necessary feature of a state trading enterprise in this environment. Reviewing the effectiveness of the International Wheat Agreement in the mid-1950s, Farnsworth (1956: 240) pointed to a 'labyrinth of national price and trade controls that existed during 1949-56'. She argues that the apparent success of the IWA in achieving price stability was more a function of the post-war wheat market, referring particularly to Canadian export pricing policy in the period 1935–1956. Gordon-Ashworth (1984: 145) agrees with this analysis, arguing that the price stability of the 1950s 'was not wholly nor even principally attributable to the workings of the Agreement itself', referring to North America as the location of 'the balance of power in international wheat transactions' (144).

On balance, the IWA was a 'boon' to importers (O'Connor 1982: 241) and they probably benefited more from the arrangement than did the exporting parties. McPherson (1949: 4) suggested that the exporters gave 'an immediate price concession for future security', seeing this as 'good business because of the trend in wheat production'. Some Australian wheat growers were less enthusiastic. They were concerned that growers had not been consulted in the process of negotiating the agreement and had been presented with a 'fait accompli' which was 'one-sided' in favour of importers (Nock 1949: 5). Contemporaneous discussion of the wheat market was tempered by a desire to avoid the difficulties of the 1930s and the International Wheat Agreement was justified in these terms. Dunsdorfs (1956: 341) argues that in his opinion, 'the participation of Australia as a major producer of wheat in the I.W.A can be accepted. It is a factor which brings stability into the industry and makes an end of the difficulties encountered in the thirties'.

The 1948 Australian Wheat Board was therefore established in an environment, both domestically and internationally, in which ideas of government intervention in commodity markets to address price instability were strongly held. The experiences of the early 1930s had convinced governments, farmers, and economists alike that some form of control was required over the market for wheat to provide price stability for producers and some assurance of supply for countries dependent on imports. Pierson has criticised the historical institutionalist literature for too quickly assuming that an institution's creators act instrumentally, expecting a particular institution to deliver particular outcomes (Pierson 2000: 478). It is clear however, that the Australian Wheat Board, the Canadian Wheat Board and the International Wheat Agreements are examples of institutions established with a very strong means-end logic. Governments either individually or collectively acted to address a very real problem by seeking to take control of a wheat market that was delivering unacceptably high levels of price volatility.

Conclusion

This chapter has described the birth of the Australian Wheat Board in 1948. The Board embodied the triumph of the values of agrarian collectivism over the concerns of the wheat traders and the supporters of the free market. By 1948 the Australian Wheatgrowers Federation had consolidated into a more professional and influential organisation than it had been at its formation in 1931 and the path was set for six decades of government involvement in wheat marketing in Australia. Dunsdorfs suggests that this was an important achievement, noting that 'out of the greatest calamity the wheat-growers emerged as a politically more powerful group than ever in their history' (Dunsdorfs 1956: 343). The Wheat Board was also born into an international institutional environment that favoured government intervention in the market for wheat, driven largely by memories of the hardships of the 1930s.

Although it began life in very similar circumstances to the Australian Wheat Board, the Canadian Wheat Board has not been subject to the same level of major institutional change. Those considering change in Canada, and there have been several attempts in recent years, may wish to ponder the evolution, and eventual demise, of the Australian arrangements and consider the lessons to be learnt from its story. The next chapter discusses the gradual demise of 'orderly marketing' as a policy approach in Australia and the corresponding evolution of wheat marketing arrangements towards their first major institutional shock.

References

Andrews J (1936) The present situation in the wheat-growing industry in southeastern Australia. Econ Geogr 12(2):109–135

Australian Labor Party (2009) History of the Australian Labor Party. Official website of the Australian Labor Party. http://www.alp.org.au/about/history.php. Accessed 15 December 2009

Australian Wheatgrowers Federation (1931) Minutes of Australian Wheatgrowers' Federation held in Melbourne on February 26th 1931. Noel Butlin Archives Centre N83, Box 37, Canberra

Botterill LC (2005) Policy change and network termination: the role of farm groups in agricultural policy making in Australia. Aust J Polit Sci 40(2):1–13

Britnell GE, Fowke VC (1949) Development of wheat marketing in Canada. J Farm Econ 31 (4, Part 1):627–642

Daugbjerg C (1999) Reforming the CAP: policy networks and broader institutional structures. J Common Mark Stud 37(3):407–28

Davey P (2006) The nationals: the progressive, country and national party in New South Wales 1919 to 2006. The Federation Press, Sydney

Dunsdorfs E (1956) The Australian wheat-growing industry 1788-1948. The University Press, Melbourne

Farnsworth H (1956) International wheat agreements and problems, 1949-56. Q J Econ 70(2):217–248

Fowke VC (1947) Canadian agricultural policy: the historical pattern. University of Toronto Press, Toronto

Furtan WH (2005) Transformative change in agriculture: The Canadian Wheat Board. Estey Cent J Int Law Trade Policy 6(2):95–107

Gabb J, MP (1930) Wheat Marketing Bill 1930: Second Reading Debate. Parliamentary Debates, 15 May 1930

Golay FH (1950) The International Wheat Agreement of 1949. Q J Econ 64(3):442–463

Gordon-Ashworth F (1984) International commodity control: a contemporary history and appraisal. Croom Helm, London/Canberra

Grant W, MacNamara A (1995) When policy communities intersect: the case of agriculture and banking. Political Studies XLIII(3):509–515

Harbury CD (1954) An experiment in commodity control - the International Wheat Agreement, 1949-53. Oxf Econ Pap 6(1):82–97

Harper CW (1928) Digest of paper on the collective marketing of Australian produce, with special reference to wheat. Econ Rec 4(Supplement):46–50

Hodan M (1954) Economic aspects of the International Wheat Agreement of 1949. Econ Rec 30(1–2):225–231

International Wheat Council (1991) The International Wheat Council and the International Wheat Agreement. International Wheat Council, London

Latham J, the Hon (1930) Wheat Marketing Bill 1930: Second Reading Debate. Parliamentary Debates, 2 May 1930

MacGibbon DA (1952) The Canadian grain trade 1931-1951. University of Toronto Press, Toronto

Mahoney J (2000) Path dependence in historical sociology. Theor Soc 29:507–548

Marsh D, Rhodes RAW (1992) Policy communities and issue networks: beyond typology. In: Marsh D, Rhodes RAW (eds) Policy networks in British government. Clarendon Press, Oxford

McPherson EE (1949) Conditions and benefits. In: Commonwealth Bank of Australia (ed) The International Wheat Agreement. Commonwealth Bank of Australia, Sydney

Mitchell G (1969) Growers in action: official history of the Victorian Wheat and Woolgrowers' Association, 1927-1968. The Hawthorn Press, Melbourne

Morriss WE (1987) Chosen instrument: a history of the Canadian Wheat Board, the McIver years. Reidmore Books, Edmonton

Nock HK (1949) A criticism of the conditions. In: Commonwealth Bank of Australia (ed) The International Wheat Agreement. Commonwealth Bank of Australia, Sydney

O'Connor C (1982) Going against the grain: the regulation of the international wheat trade from 1933 to the 1980 Soviet Grain Embargo. Boston Coll Int Comp Law Rev 5(1):225–270

Patton HS (1937) Observations on Canadian Wheat Policy since the World War. Can J Econ Polit Sci 3(2):218–233

Peters BG (1999) Institutional theory in political science: the 'new institutionalism'. Pinter, London

Peterson J, Bomberg E (1999) Decision-making in the European Union. St Martin's Press, New York

Pierson P (2000) The limits of design: explaining institutional origins and change. Governance 13(4):475–499

Royal Commission on the Wheat Flour and Bread Industries (1934) First report. Government Printer, Canberra, 2 August 1934

Royal Commission on the Wheat Flour and Bread Industries (1935) Second report. Government Printer, Canberra

Sartori G (1990) A typology of party systems. In: Mair P (ed) The West European party system. Oxford University Press, Oxford

Schmitz A, McCalla A (1979) The Canadian Wheat Board. In: Hoos S (ed) Agricultural marketing boards - an international perspective. Ballinger Publishing Company, Cambridge

Smith RFI (1969) 'Organise or be damned': Australian Wheatgrowers' Organisations and wheat marketing, 1927-1948. PhD Dissertation, Political Science, Research School of Social Sciences, The Australian National University

Smith RFI (1974) The Scullin government and the wheatgrowers. Labour Hist 26:49–64

Smith MJ (1990) The politics of agricultural support in Britain: the development of the agricultural policy community. Dartmouth, Aldershot

Smith MJ (1992) The agricultural policy community: maintaining a closed relationship. In: Marsh D, Rhodes RAW (eds) Policy networks in British government. Clarendon, Oxford

Thelen K (1999) Historical institutionalism in comparative politics. Annu Rev Polit Sci 2:369–404

Thelen K, Steinmo S (1992) Historical institutionalism in comparative politics. In: Steinmo S, Thelen K, Longstreth F (eds) Structuring politics: historical institutionalism in comparative analysis. Cambridge University Press, Cambridge

Tyszynski H (1949) Economics of the Wheat Agreement. Economica 16(61):27–39

Wadham SM (1944) Reconstruction and the primary industries, vol 7, Realities of reconstruction. Melbourne University Press, Melbourne

Watson AS, Parish RM (1982) Marketing agricultural products. In: Williams DB (ed) Agriculture in the Australian economy, Second edn. Sydney University Press, Sydney

Whitwell G (1993) Regulation and deregulation of the Australian wheat industry: the 'great debates' in historical perspective. Aust Econ Hist Rev XXXIII(March):22–48

Whitwell G, Sydenham D (1991) A shared harvest: the Australian wheat industry, 1939-1989. Macmillan Australia, South Melbourne

WTO (World Trade Organization) (2008) Revised draft modalities for agriculture. TN/AG/W/4/Rev.4 Committee on agriculture special session. 6 December 2008

Chapter 4
From Orderly Marketing to Deregulation 1948–1988

Keywords Wheat industry developments 1948–1988 • Australia • Institutional change • Rural policy development • Australian Wheat Board

The *Wheat Stabilization Act 1948* was renewed on a regular basis throughout the 1950s and 1960s. This was a period of entrenchment of the marketing arrangements (Cockfield and Botterill 2007), with little change to the nature of the support offered to growers through the stabilisation provisions and with continued government involvement in the operations of the Australian Wheat Board. The economic environment within which the Board was operating was changing, however, and alternative approaches to rural policy began to be raised in public policy debate.

In the 1970s and the 1980s, Australia, in common with much of the developed world, experienced a major shift in economic philosophy. Keynesian economics was rejected and an increasingly neoliberal approach adopted. The shift in economic thinking triggered a major opening up of the Australian economy through a reduction in protection, deregulation of financial markets, the floating of the Australian dollar and an enthusiastic engagement in the Uruguay Round of multilateral trade negotiations. Government enterprises, such as the Commonwealth Bank, were privatised. This ideational shift did not bypass agriculture, or its venerable institutions. The Australian Wheat Board's operations came under scrutiny from the Industries Assistance Commission, government reports and, indirectly, a Royal Commission.

This chapter tracks important changes in the rural policy paradigm in Australia and examines the implications for collective wheat marketing. The Australian Wheat Board was not immune to challenge, but, like all institutions, it adapted to its changing environment and adopted various strategies of institutional reproduction which allowed it to protect its core value of agrarian collectivism; embodied in its monopoly over wheat marketing and grower involvement in the setting of wheat industry policy. An important part of its defence strategy relied on the growing closeness between the Board and the Australian Wheatgrowers Federation (AWF).

L.C. Botterill, *Wheat Marketing in Transition: The Transformation of the Australian Wheat Board*, Environment & Policy 53, DOI 10.1007/978-94-007-2804-2_4,

Once an institution becomes established, others tend to evolve around it, either embodying similar values and providing a support network, as was the case of the AWF, or by providing an alternative venue for conflicting values to be heard (Daugbjerg and Botterill 2011).

While the AWF provided a complementary institution which was heavily 'invested' (Thelen 1999: 391) in the values at the heart of the wheat marketing arrangements, an important structural change took place in the broader agricultural policy community in 1979. After decades of division among agricultural producer organisations, the National Farmers Federation (NFF) was formed. It adopted with enthusiasm the new economic paradigm of market liberalism and the agricultural policy community became tighter, more cohesive and more focused as the policy approach of the peak farm body became more congruent with that of the government. However, the Australian Wheatgrowers Federation (later renamed the Grains Council of Australia) remained as a commodity council within the NFF and continued to pursue its policy of support for collective wheat marketing. The new policy paradigm required some adaptation but this was able to be accommodated within the existing institutional structure which continued to enjoy broad political and public support.

A Changing Rural Policy Paradigm

By the mid-1960s, it was becoming increasingly clear that the highly interventionist agricultural policy settings that had been in place in Australia were unsustainable. Voices of opposition to the high level of government involvement in agricultural markets began to be heard. Agricultural economists pointed to the costs and inefficiencies arising from the various policy settings and bemoaned the lack of economic literacy among farm leaders and rural politicians (see for example Lloyd 1970; Makeham and Bird 1969; Mauldon 1968; McKay 1965, 1967). Some Liberal politicians, notably Bert Kelly and John Hyde, themselves graingrowers, began to voice opposition to the existing arrangements.

A change of government in 1972 brought with it opportunities to revisit the policy settings. Two key policy innovations opened the way for a more liberalised rural policy. The first was the establishment of the Industries Assistance Commission and the second was the commissioning by the incoming Labor government of a Rural Policy Green Paper.

The Industries Assistance Commission (IAC) was set up in 1974, replacing an earlier body, the Tariff Board, which had been established as a permanent body by legislation in 1924. In its early years, the focus of the Tariff Board was on the protection of 'economic and efficient' industries, although neither of these terms was adequately defined. The last Chairman of the Tariff Board, Alf Rattigan described the Board as 'an *ad hoc* operator' with its work 'generally initiated by requests from Australian manufacturers for increased duties for particular products' (1986: 22). Although the Board's legislation allowed for the examination of assistance to primary industries, its focus was on the protection of manufacturing. Rattigan considered that the Board's approach resulted 'in a creeping increase in the level of tariff protection' (23) in Australia. By the 1960s, just as economists were questioning the levels of support for agriculture, members of the Tariff Board similarly began questioning the support being provided to manufacturing. In its 1966–1967 Annual Report the Tariff Board proposed a change in its approach to assessing industry assistance, suggesting that it commence a systematic review of the tariff. This would include increased public scrutiny of assistance which had previously been developed piecemeal and out of public view. This proposal was not well received by the government and for the rest of the 1960s, the Tariff Board and the government were in ongoing dispute over the Board's role.

On winning government in 1972, the Labor Party sought a review of industry assistance and the establishment of an independent Commission to provide advice to government on support to both primary and secondary industries. An eminent economist, Sir John Crawford was selected to advise the government on how such a Commission might be set up and operate. In his report, Crawford identified three key reasons for establishing the Commission:

- To 'assist the government to develop co-ordinated policies for improving resource allocation';
- To 'provide advice on those policies which is disinterested'; and
- To 'facilitate public scrutiny of those policies'. (Crawford 1973: 34)

Following Crawford's report, the government established the Industries Assistance Commission, marking a significant change to the way in which Australian industry policy was made. In contrast to the *ad hoc* approach that Rattigan had noted, the government introduced a systematic examination of requests for industry assistance. The Prime Minister told the Parliament that 'The first and most important reason for establishing the Commission is to allow public scrutiny of the process whereby governments decide how much assistance to give to different industries' (Whitlam 1973: 1632). He further stated that 'Government shall not take any action to provide assistance to a particular industry until it has received a report on the matter from the Commission' (Whitlam 1973: 1633). This level of transparency and public inquiry drew criticism from the farmers' political representatives, the Country Party. The Party's leader protested that 'what this means, of course, is the end of the long-established and successful system under which industry policy has been devised – the system of discussion, consultation and negotiation between industry and government' (Anthony 1973: 2356). It is worth noting that the Country Party's coalition partner, the Liberal Party crossed the floor to vote with the government to pass the IAC legislation; one of the few occasions on which the parties have split publically over a major policy issue.

For rural industries the important development, and likely a large part of the reason for the Country Party's opposition to the Industries Assistance Commission, was the inclusion of assistance to primary industries in the IAC's ambit. As Warhurst (1982: 25) has noted, 'primary industry inquiries quickly became a significant call on IAC resources. Of the 18 references received by the Commission in the first 6 months of its existence, 9 related to rural industries'. The IAC held inquiries into assistance to the wheat industry in 1978, 1983 and 1988 as well as a broader inquiry into agricultural statutory marketing arrangements in 1991. The recommendations of these inquiries and the Government's response are discussed below.

The second development which signalled a liberalising trend was the overall policy direction proposed in the 'Principles and Guidelines' in the Rural Policy Green Paper (Harris et al. 1974). These were more consistent with the policy suggestions of the agricultural economists than the policy settings that were in place at the time. It is worth noting that as well as his role in recommending the parameters for the new IAC, Sir John Crawford was also a member of the Working Group which developed the Green Paper. The Green Paper signalled an important shift away from the prevailing rural policy philosophy of intervention, dealing 'tactfully and persuasively with many rural shibboleths' (Lloyd 1982: 364). Among its principles, the paper argued that

> A basic postulate of rural policy is that the market is generally the most effective method of allocating productive resources, but that governments will not accept the full implications of the free working of the market mechanism in the agricultural sector and recognise the need to intervene to improve the manner in which the market operates or to compensate for its consequences. (Harris et al. 1974: 48)

Although this statement implies continued intervention by government in the markets for primary production, the report clearly favoured the facilitation of market-driven adjustment in agriculture rather than intervention to mask market signals. The Green Paper included a chapter on stabilisation which reflected this concern. It argued that 'If market prices are prevented from effectively performing their allocative function, significant social costs can result' (Harris et al. 1974: 60). It also pointed out that stabilisation and equalisation schemes 'may have the effect of reducing export earnings over the long run from the resources employed'. As an example the report cited wheat prices and acreages in the period 1948–1949 to 1952–1953 which it argued 'were probably lower … than they would have been in the absence of stabilisation – at a time when the world was 'crying out' for wheat and willing to pay very high prices for it' (Harris et al. 1974: 60). More directly, the report concluded that for the periods it examined 'the wheat stabilisation scheme has had a very minor effect either on prices or gross incomes from wheat'. It was more positive about other elements of the wheat marketing arrangements, arguing that

> other government intervention in wheat marketing (apart from stabilisation) has probably contributed to the low variability of prices shown here. Particularly relevant are the Wheat Board's roles as a large seller on export markets and the international pricing arrangements for wheat. These forms of government action, as much as stabilisation, have cushioned the large prices falls that are of particular concern to producers. (Harris et al. 1974: 66)

The chapter on marketing arrangements in the Green Paper was ambivalent towards government intervention, reflected in its conclusion that 'Statutory marketing boards may be an effective means of improving farmers' bargaining power, to obtain economies in handling or merchandising, or to stabilise prices; however, marketing boards are themselves capable of inertia and inefficiency particularly in a non-competitive environment' (Harris et al. 1974: 179). The Green Paper did not offer a direct challenge to the values underpinning collective wheat marketing, but its overall message was one of a preference for allowing market signals to influence production and resource allocation decisions. This in itself marked a shift away from the approach seen in the first half of the twentieth century.

The changing policy mood and the increased scrutiny by the Industries Assistance Commission of assistance to both primary and secondary industries provided an important impetus for change in the representation of primary industry interests. As noted, after decades of fragmentation and internal division, farm organisations in Australia finally united in 1979 to form the National Farmers Federation (NFF). Farmers saw the Industries Assistance Commission inquiry process as an important avenue for reducing assistance to manufacturing industries and thereby reducing their input costs. Although the farmers' political representatives, the Country Party, had supported the 'protection all round' approach of the post war era, farm groups had not always agreed with this policy stance. When the Industries Assistance Commission legislation was being debated, the Country Party was highly critical, suggesting that the Commission would be too powerful, undemocratic and was a step along the path towards central planning. The Country Party's spokesman told the Parliament that 'This proposed body should be called, not the Industries Assistance Commission, but the Industries Assistance-Withdrawal Commission' (Anthony 1973: 2354). Rattigan argues that this position was inconsistent with the attitudes of farmers themselves as 'none of the primary industry organizations who represented most of the rank and file of the Country Party opposed the creation of the IAC' (Rattigan 1986: 190). When the Country Party was returned to government in 1975 in coalition with the Liberal Party, farmers were opposed to the new Minister for Primary Industry's suggestion that responsibility for assistance to primary industry be removed from the IAC (Rattigan 1986: 264).

Beyond agriculture, reform of the Australian economy accelerated during the 1980s, particularly following the election in 1983 of the Hawke Labor Government. The Australian financial system was deregulated, the dollar was floated, a program of privatisation began and industry assistance was steadily dismantled. Australia became active in the Uruguay Round of multilateral trade negotiations that commenced in 1986. In its position as leader of the Cairns Group it championed the cause of free trade in agriculture. As a social democratic party, the Australian Labor Party has strong ties to the Trade Union movement and it used that relationship to implement incremental, but ultimately significant, changes to the labour market in Australia, moving industrial relations from a long tradition of collective bargaining within a formal structure towards enterprise-based bargaining. These changes set the groundwork for a move to individual contracts introduced in the 1990s by its Liberal Party successor. Paul Kelly (1994) has described the period of the 1980s as the 'end of certainty' as the structures that had underpinned the formation of the Australian federation were gradually dismantled and replaced with market-based approaches to economic management, including for agriculture.

Institutional Development and Change

The wheat industry was not immune from the liberalising trend in Australian economic policy from the 1960s onwards. However, change was highly incremental and, by comparison with the deregulation of other areas of agricultural marketing, relatively slow. Although the transformations of the mid-1990s described in the next chapter were dramatic, they built on years of gradual shifts in the focus of government intervention in Australia's wheat marketing arrangements. These were part of a progressive move away from protection of wheat growers from price and income volatility to an increased exposure to world prices and market signals. The shift from the marketing of wheat of 'Fair Average Quality' to the operation by the Australian Wheat Board of differentiated pools and payment for particular grain characteristics occurred gradually over many years. This constituted a move away from the collective approach to wheat marketing which spread costs and risks across all growers and which had underpinned the establishment of the wheat stabilisation scheme in 1948. The emerging approach was more market-oriented and individualistic, no longer treating Australian wheat as a homogeneous product.

The first major change to the wheat stabilisation plans was the imposition of quotas in 1969. The quotas were in response to potential over-production of wheat and concern that the Wheat Board would not be able to dispose of the export wheat crop. Whitwell and Sydenham (1991: 187) argue that the industry agreed to the imposition of quotas partly 'to forestall mounting criticism of the wheat stabilisation scheme'. Adapting to threats by conceding minor changes to their operations, is not an unexpected tactic of institutions faced with an external challenge. By accepting small, bearable change the core values of the organisation can be protected and preserved. Whitwell and Sydenham note that 'Critics of the wheat scheme were more vocal and greater in number once the extent of the carryover for 1968–69 became obvious' (Whitwell and Sydenham 1991: 185). The quotas were not well received by growers who were unhappy with the way they were implemented. They were administratively burdensome for the Australian Wheat Board and they were inequitable between new and established growers and between larger and smaller operations (Whitwell and Sydenham 1991: 186–187). They also provided an incentive for the development of a black market trade in wheat.

The quotas were also the reason behind one of the more obscure footnotes in Australian political history – the secession from the Commonwealth of the Hutt River Province in Western Australia which claims its status as an independent sovereign state. It issues its own currency, knighthoods and university qualifications. The Province, which later became a principality, was formed in direct response to the Western Australian government's administration of the wheat quota scheme.[1]

[1] The Australian government does not recognise the secession and considers the principality to be a private business.

The second change which altered the nature of the wheat marketing scheme was the abandonment of the FAQ – Fair Average Quality – classification of wheat and early steps towards segregation of the wheat crop on the basis of quality and milling characteristics. The Royal Commission on the Wheat, Bread and Flour Industries described the FAQ system as follows:

> Under this system a composite sample is made up each year from the crop in each district and this is declared to be the f.a.q. for the State for that particular year. The declaration with an attendant sample is sent to the Baltic exchange in London and forms the basis of dealings in that season's wheat for that State. (Royal Commission on the Wheat Flour and Bread Industries 1935: 168)

The FAQ system reflected the collective sentiment underpinning the original wheat stabilisation schemes by implying that all wheat growers produced an equally valuable product. The system also ensured that farmers in small rural communities knew that they were receiving the same payment per tonne for their crop as their neighbours; a feature of the system that was considered desirable in close-knit communities in the wheat belt. In 1967 several different FAQ grades were introduced but these were not particularly helpful to buyers. As Watson and Parish (1982: 345) argue, the FAQ grading system 'was perhaps the most self-effacing product description yet devised'. In 1974, Fair Average Quality was replaced by Australian Standard White (ASW). This was still a large class but had the advantage of emphasising an important characteristic of Australian wheat – its whiteness (Whitwell and Sydenham 1991: 272).

The Industries Assistance Commission Reports

The first real institutional challenge to the wheat stabilisation arrangements came in 1978 with the first of the three Industries Assistance Commission reports into wheat marketing. The Commission was blunt in its assessment. In its view, and a point foreshadowed in the 1974 Green Paper analysis of wheat price movements, the stabilisation arrangements had had 'little impact on the stability of prices or incomes received by wheatgrowers'. The Commission recommended that 'the traditional form of stabilization of returns for wheat be discontinued' (IAC 1978: i). The Commission also recommended that the Australian Wheat Board's monopoly as sole seller of wheat on the domestic market be discontinued but, although it could not assess the alleged advantages of the export monopoly, it recommended that the Board retain its sole exporter status (IAC 1978: v). The Commission noted that the Australian Wheat Board had been using private traders for the export of wheat to some markets and recommended that this practice continue. The IAC also recommended that, instead of borrowing from the Rural Credits Department of Australia's central bank, the Reserve Bank, to finance harvest payments, the Wheat Board should access commercial finance, with some finance to be either provided or guaranteed by the Commonwealth government. As the IAC itself later noted of the 1978 report, 'The Government rejected the majority of the Commission's recommendations' (IAC 1988: 26), including the proposal to deregulate the domestic wheat market.

In response to the IAC report, the Government allowed the Board to offer different prices for wheat depending on its end use and moved from stabilisation funds to government underwriting of pool returns in the form of a Guaranteed Minimum Price (GMP). Under this arrangement, the Government payments to the Australian Wheat Board were triggered when the net return for a particular wheat pool was less than the GMP for that pool, a similar approach to that taken by the Canadian Wheat Board in its early days but without the option for growers to sell into the free market when prices were high. A further change was made which allowed for the direct negotiation between growers and buyers of particular deliveries of wheat, however they were still required to be included in the Board's pooling arrangements and growers were charged a fee for storage and handling, even if these services were not used. The introduction of the grower-to-buyer option was only a limited concession. In 1980–1981 only 6,166 tonnes of wheat were delivered under these provisions and it was clear that a black market had emerged in directly-traded wheat (IAC 1983: 21–22). There was also a 'grey' market in wheat traded across state borders; not strictly illegal due to the nature of the Australian constitution but certainly against the spirit of the wheat stabilisation scheme.

The next Industries Assistance Commission inquiry into wheat marketing was undertaken in 1983 (IAC 1983). The Commission once again recommended that the domestic market for wheat be opened to competition and that the price no longer be administered. The Commission was concerned that the Wheat Board's domestic monopoly was distorting the market for grain handling and transportation, thereby increasing the costs to growers. The introduction of competition, it argued, would encourage the bulk handling authorities to pay 'greater attention' to pricing policies (IAC 1983: 40). On the issue of segregation the Commission pointed to the cross subsidisation by growers of high quality wheat of those producing poorer quality grain. It recommended that the Board establish a greater number of pools for payments to growers.

The Government's response, *The Wheat Marketing Act 1984*, made some concessions to a freer domestic market for wheat by allowing the sale of stockfeed wheat outside the pooling arrangements and under a permit issued by the Australian Wheat Board. It introduced some scope for differentiating between categories of wheat and gave the Australian Wheat Board power to operate on the futures market. It also made changes to the composition of the Board, moving from a representative board to one which included greater expertise in finance and marketing. In introducing the legislation, the Minister for Primary Industries and Energy noted that the 'functions of the AWB have dramatically changed in recent years' (Kerin 1984). In addition to a grower member from each State and a grower Chairman, the Act allowed for the government to appoint up to six members of the Board, up to five of which could also be growers selected by the Australian Wheatgrowers Federation. The AWF's role in the operation of the wheat marketing system was further entrenched by the inclusion of a statutory obligation on the Wheat Board to consult with the organisation.

The IAC again examined the wheat industry in 1988 and again called for the deregulation of the domestic wheat industry. The Commission proposed the replacement of the guaranteed minimum price with a government guarantee on Wheat Board borrowings to finance the first advance payment which growers received on delivery of their wheat to the pool. This recommendation followed the triggering of the GMP underwriting system in 1986–1987 at a cost to the Commonwealth Budget in excess of $200 million (IAC 1988: 12). The Commission was also concerned that the arrangements had 'the potential to influence the allocation of resources … because they reduce[d] price uncertainty and encourage[d] higher levels of output' (IAC 1988: 84).

The Industries Assistance Commission went one step further than its earlier reports and questioned the value of the export monopoly by proposing that the monopoly only apply to markets 'in which price premiums could be eroded by multiple sellers of Australian wheat' (IAC 1988: 7). These premium markets would be 'prescribed by the Minister, after consultation with the AWB, the trade and other appropriate government departments' (IAC 1988: 7). The Australian Wheat Board was already using international grain traders to dispose of a sizeable portion of the wheat crop, causing the Commission to observe that

> profit-taking by traders does not necessarily imply that growers would be worse off. It is quite plausible for trading organisations to make profits and, at the same time, for growers to achieve higher returns than those which they might attain from selling to a central marketing agency. (IAC 1988: 112)

The Commission proposed some transitional or 'middle ground' alternatives to the removal of the export monopoly which addressed some of the inefficiencies it had identified with the existing system and improved the transmission of market signals to growers.

The McColl Royal Commission

Also important to the story of the deregulation of the domestic market for wheat was the Royal Commission into Grain Storage, Handling and Transport, chaired by Jim McColl, which reported in 1988. This inquiry ran in parallel with the 1988 IAC wheat inquiry. The Royal Commission noted that it was 'the first inquiry covering all aspects of grain distribution on a national basis' (Australia. Royal Commission into Grain Storage 1988: 10). As noted in Chap. 3, the Wheat Board's monopoly on grains sales had been accompanied by the establishment of monopoly handlers of wheat in each of the wheat-producing states. Under the Australian Wheat Board's legislation it was required to appoint a sole receiver of grain in each state. These receivers were owned by the State governments and their operations were underpinned by State legislation which, among other things, restricted the use of road transport for the movement of grain, thereby forcing grain on to State-run rail networks. Storage and handling costs were pooled and growers charged an average price and, following the slight easing of domestic marketing arrangements discussed above, these charges were also levied on grower-to-buyer sales and permit wheat transactions – even though the services were not used in these circumstances.

The Royal Commission's recommendations were consistent with the series of IAC reports in that they challenged the collective values underpinning the orderly marketing arrangements. The Commission pointed to both the distributional and efficiency consequences of cost pooling (Australia. Royal Commission into Grain Storage 1988: 56) and recommended that marketers should pass back to growers the 'actual charges incurred by that grower' for grain distribution (p. 136). The Royal Commission found that grain storage, handling and transport did 'not meet the criteria of economic efficiency, cost effectiveness and integration' set out in its terms of reference (p. xxv) and that cost savings of an average $9 per tonne could be attained through greater competition (p. 90).

The Commission considered the option of deregulating transport, storage and handling but opted instead to recommend a 'mixed deregulatory/regulatory approach' (Australia. Royal Commission into Grain Storage 1988: xxix). It rejected complete deregulation on the grounds of the potential for the emergence of monopoly power, concerns about the capacity of a deregulated market to ensure high standards of grain hygiene, and problems of externalities such as damage to roads and consequent inadequate road funding (p. xxix). Concerns about monopoly power and grain hygiene standards were to feature in debate some 20 years later when the Wheat Board's privatised successor was facing the loss of its export monopoly. The Commission's report suggested that State governments consider privatising their grain handling authorities, with provision for growers to share in any new ownership structure. In the absence of privatisation, the report was clear that the authorities should be restructured along corporate lines (Australia. Royal Commission into Grain Storage 1988: 146).

The Commonwealth Government strongly endorsed the Royal Commission's findings and acted in 1988 by amending the wheat marketing legislation to remove the sole receiver requirement and allow the Australian Wheat Board to 'contract with whomever will provide the best returns for growers' and to provide disaggregated charges for storage and handling where possible (Howe 1989). Given that the bulk handling authorities were established under State legislation, responsibility for much of the response to the Royal Commission's report lay with State governments. Over the course of the next several years, State governments moved to restructure their bulk handling authorities with the creation of Grainco in Queensland in 1991, Graincorp in NSW in 1992 and VicGrain in Victoria in 1995. South Australia and Western Australia retained cooperative ownership models for their bulk handlers. Graincorp listed on the Australian stock exchange in 1998, merged with VicGrain in 2000 and took over Grainco in 2003 after the latter listed on the stock exchange in 2002. South Australia's bulk handler has subsequently been taken over by ABB Ltd, the privatised former statutory Australian Barley Board.

Reviews of Statutory Marketing

As well as being explicitly considered as part of the Industries Assistance Commission's work program, the Wheat Board was impacted by the Government's 1986 White Paper *Reform of Commonwealth Primary Industry Statutory Marketing Authorities*. This paper attempted to improve the effectiveness and efficiency of the Statutory Marketing Authorities (SMAs) by introducing corporate principles into their management. This was to be achieved through the reform of Boards by basing their membership on expertise rather than on the representation of particular constituencies and through improving the accountability of the Authorities both to the industries on whose behalf they operated and to the Commonwealth Parliament. The Report noted that

> The SMAs were established at the request of producers for statutory organisations to coordinate the marketing of their produce. They are essentially commercial bodies with trading, promotional and/or regulatory powers and are funded through compulsory levies paid by growers or from the sale of compulsorily acquired produce as in the case of the AWB. (Australia. Department of Primary Industry 1986: 3)

The report emphasised that the SMAs should not have a role in agri-politics and also argued that it was inappropriate for the President or other executive members of industry associations to whom the Authority was accountable to sit on the Board (Australia. Department of Primary Industry 1986: 5–6). In order to ensure high calibre people were selected for the Board, the Government sought the establishment of 'objective selection procedures' to be overseen by a statutory selection committee (Australia. Department of Primary Industry 1986: 26). While the language of the White Paper implied the development of an arms length relationship between the SMA Boards and industry organisations, in practice the policy further embedded the Australian Wheatgrowers Federation's successor, the Grains Council of Australia (GCA) into the Australian Wheat Board's institutional structures and did little to alter the career paths of grains agri-politicians. The legislation simply required their resignation from the GCA Executive before they joined the Board, rather than allowing them to serve on the two simultaneously. The Government's report on Statutory Marketing Authorities had linked the industry function of the primary industry statutory marketing authorities with the need for industry to be involved in the selection of the majority of the Board members, specifying that 'industry representatives to selection committees will be nominated by the nationally recognised industry body/bodies' (Australia. Department of Primary Industry 1986: 20). In the case of the Australian Wheat Board, this was the Grains Council of Australia.

In 1990 a Committee of Inquiry looked at the implementation of the 1986 review of statutory marketing. The Committee pointed to the 'substantial' indirect costs to the community of statutory marketing arrangements:

> There are two dimensions to these costs. The first is the direct cost of departmental, ministerial and parliamentary supervision. The second is the indirect cost that an SMA may cause if it blocks the emergence of competing organisations which would provide marketing services in a more efficient way. (Commonwealth of Australia 1990: 12)

The Inquiry argued that SMAs therefore needed to demonstrate a public benefit arising from their existence over and above the direct benefit to their levy payers (Commonwealth of Australia 1990: 12). This argument about net benefit to the broader public was reinforced in the National Competition Policy Inquiry which reported in 1993 (Hilmer et al. 1993).

The 1990 committee of inquiry also addressed the issue of industry representation. It argued that

> The arrangement where an SMA reports to a representative peak industry body can disenfranchise individual levy payers if they are not members of that body and do not wish to join, or if the body's policy is contrary to their own views. (Commonwealth of Australia 1990: 22)

However, it concluded that because the Board structure had changed from representational to expertise-based as a result of the 1986 White Paper, the 'question of producer control of boards through representation does not appear to be an issue now' (Commonwealth of Australia 1990: 29). This point is debatable as, although members of the Grains Council Executive could not also be on the Australian Wheat Board, there was nothing to prevent their nominating to be on the Board and then resigning from the GCA Executive. Although this removed some potential conflict of interest, it preserved the close linkages between the Board and the Grains Council of Australia and strengthened the perception of the career path of grains industry agri-politicians from GCA to positions on the Australian Wheat Board.

In 1991 the Industries Assistance Commission's successor, the Industry Commission, released a report on *Statutory Marketing Arrangements for Primary Products*. The purpose of this report was to provide 'an 'in-principle' examination of the central issues' associated with statutory marketing, viz

- the objectives of statutory marketing arrangements;
- their economic effects;
- ways to improve their efficiency; and
- priorities for change. (IC 1991: 1)

The Inquiry's Terms of Reference focused on possible inefficient resource use arising from regulatory arrangements; it was not tasked with looking at arrangements for specific commodities. Of relevance to the wheat industry, the report identified problems with the collective nature of pooling and compulsory acquisition arrangements. It argued that aspects of the arrangements catered 'for the average – or even the lowest common denominator – producer or purchase' (IC 1991: 3) and led to distorted market signals. Pooling was acknowledged as spreading risk among growers but was also seen as a disincentive to the development of efficient management. The Report pointed to the exemption of SMA activity from anti-competitive provisions of trade practices legislation. It also called on governments to 'review their procedures for deregulation and privatisation of SMAs (or features of their operations)' (IC 1991: 115).

The Changing Face of the Rural Policy Community

In 1979 there was an important development in the rural policy community. As noted, the fractured and disorganised industry associations that had characterised the rural industries in Australia finally agreed to amalgamate into a single peak body (Connors 1996). This followed several false starts in the 1970s which had been stymied by serious policy differences between export-oriented farm groups that favoured free trade and deregulated markets, and the industries which produced for the domestic market and were content with price supports and high levels of government intervention. Apart from differences between industries, there were disputes within industries over policy direction; these disputes had resulted in competing groups claiming to represent the industry. The wheat industry was one of those which was divided.

For many years, an important cultural divide existed in Australian agriculture which was reflected in the various competing industry associations. This is generally characterised as being a division between 'farmers' and 'graziers', although both groups grow crops and raise livestock. The graziers are seen as the silvertails of the rural sector; they generally farm larger holdings, their children are more likely to attend private schools, their wives are less likely to work on the farm and they are generally supporters of free trade in agriculture. They are seen as the descendents of the 'squattocracy' of the early days of Australian settlement. By contrast, farmers are the descendents of the 'industrious yeomanry' that colonial governments sought to encourage through land reforms in the nineteenth century and through closer settlement and soldier settlement schemes. The wives of 'farmers' are more likely to be active partners in the farm operation and as a group, farmers are more inclined to support government intervention in the markets for agricultural products. These cultural differences spilled over into the industry associations within the rural sector. As Connors (1996: 22) points out

> They remained divided for almost 90 years as the original battle over land ownership gave way to bitter contests over marketing farm produce, with farmers demanding government intervention and graziers adhering to the free market. The gulf between the two groups was widened by differences in wealth, property size, education levels and social status.

The formation of the National Farmers Federation was therefore no mean feat. Observers keenly watched to see which of the two groups, the farmers or the graziers, would dominate the new Executive of the Federation and therefore the policy direction of the new peak body. When the newly elected Executive appeared to be dominated by farmers, newspaper reports suggested that small operators would hold the power within the organisation and thus determine policy direction (Hodgkinson 1979; The Land 1979). However the grazier organisations managed to secure the top secretariat positions for their staff, thereby ensuring their economic approach prevailed. As Connors notes,

> Leading graziers saw greater benefits in pushing their former staff into senior positions on the NFF and its commodity councils than in demanding leadership posts for themselves. When full time staff serve part time executives there are opportunities for staff to have considerable influence. (Connors 1996: 214)

From the time of its formation the National Farmers Federation became one of the most enthusiastic advocates of the deregulation of the Australian economy. It saw benefits in reducing the costs of farm inputs and from increased efficiency within its own industry. In a policy paper released in 1981, the NFF argued that

> NFF does not believe that any industry — rural, mining, manufacturing, or tertiary — whether highly protected or not — should be permanently shielded from the forces of economic change. The overall interests of the economy demand that all industries must participate in the inevitable adjustment process. (National Farmers Federation 1981: 48)

The NFF's policy approach was consistent with the shifting economic approach taking place within government. The support of the peak farming lobby group and the general trend towards economically liberal policy settings across the economy was reflected in the incremental dismantling of the old system of orderly marketing.

Perhaps fortunately for the wheat industry and the defenders of collective marketing, the NFF's structure allowed policy space for advocates of orderly marketing to continue to operate. While the NFF itself took responsibility for policy on cross-sectoral issues such as trade, quarantine, drought and structural adjustment policy, it retained within its structure a collection of commodity councils affiliated with the peak body who addressed industry specific issues. In their study of how the policy process deals with conflicting values, Thacher and Rein (2004) propose a number of strategies, one of which is the construction of 'firewalls'. Under this strategy, different values are handled by different institutions and

> By focusing the attention of each institution on a subset of the values that ultimately matter to public policy, it is possible to simplify the task of practice and keep the pathologies of value conflict at bay. That arrangement helps ensure that each value has a committed defender—that no value becomes neglected because the institutions that should pursue it have become sidetracked by other concerns. (Thacher and Rein 2004: 469–470)

The structural separation within the National Farmers Federation family of organisations provided such firewalls and allowed for the management of policy differences which would otherwise have caused problems for the new body. For example, the interests of the grains industry are not always compatible with those of livestock producers, particularly lot-feeders who want access to cheap, high quality feed grains. The tension between these two groups came to a head in the mid-1990s when drought reduced the grain crop to such an extent that there were calls for imports of grain to meet the shortfall. The debate over the development of an appropriate quarantine protocol to allow such imports was protracted and bitter.

The creation of the commodity councils allowed particular industries to pursue industry-specific policy without reference to the NFF and its free market approach. This meant that the Grains Council of Australia was able to develop a close relationship with the Australian Wheat Board and to defend the wheat marketing arrangements vigorously. A specific grains industry policy community evolved comprising the Grains Branch of the Department of Primary Industries and Energy, the Australian Wheat Board and the GCA. As mentioned above, the wheat industry had been one of those in which there was more than one industry voice prior to the establishment of the NFF. Wheat producers have never been unanimous supporters of the stabilisation arrangements and neither has the export monopoly attracted universal support. Efficient producers of high quality grain were frustrated by the pooling arrangements, particularly in the days before the Wheat Board abandoned the Fair Average Quality classification. However, their voices were excluded from the policy community while the GCA's policy position was in lock step with the values underpinning the establishment of the system of collective wheat marketing. Following the release of the IAC reports and that of the Royal Commission into Grain Handling, Storage and Transport, the industry association fought against changes to the Wheat Board's role.

Conclusion

By 1988, a series of Industries Assistance Commission reports, the Royal Commission into Grain Storage, Handling, and Transport and the Government inquiry into the operation of Statutory Marketing Authorities all challenged the values underpinning the collective marketing arrangement embodied in the Australian Wheat Board and supported by the AWF/GCA. The economic paradigm within which government policy was made had undergone a major shift to free market liberalism and the new National Farmers Federation embraced this shift with enthusiasm. During this period of challenge from about the late 1960s, the Wheat Board was able to adapt through relatively minor changes to the way it did business. None of the changes could be interpreted as the type of 'critical juncture' referred to in much of the historical institutionalist literature; rather change was incremental and adaptive. All this changed in 1989 when the government of the day removed a major plank of the collective marketing system – the Wheat Board's monopoly over the domestic market.

References

Anthony D, the Hon (1973) Industries Assistance Commission Bill: Second Reading Debate. House of Representatives Hansard, 18 October 1973

Australia. Department of Primary Industry (1986) Reform of Commonwealth Primary Industry Statutory Marketing Authorities. A Government Policy Statement Australian Government Publishing Service, Canberra, January 1986

Australia. Royal Commission into Grain Storage, Handling, and Transport (1988) Royal Commission into grain storage, handling and transport. Volume 1: Report. Australian Government Publishing Service, Canberra

Cockfield G, Botterill LC (2007) Deregulating Australia's wheat trade: from the Australian Wheat Board to AWB Limited. Public Policy 2(1):44–57

Commonwealth of Australia (1990) Review of the Commonwealth Primary Industry Statutory Marketing Authorities. Report to the Minister for Primary Industries and Energy by the Review Committee Australian Government Publishing Services, Canberra

Connors T (1996) To speak with one voice. National Farmers Federation, Canberra
Crawford JG (1973) A Commission to advise on assistance to industries: report by Sir John Crawford. Australian Government Publishing Service, Canberra
Daugbjerg C, Botterill L (2011) Ethical food standard schemes and global trade: challenging the WTO? 85th Agricultural Economics Society Conference, Warwick, UK, 18–20 April
Harris SF, Crawford JG, Gruen FH, Honan ND (1974) The principles of rural policy in Australia: a discussion paper. Report to the Prime Minister by a Working Group AGPS, Canberra
Hilmer FG, Rayner MR, Taperell GQ (1993) National Competition Policy. Report by the Independent Committee of Inquiry Australian Government Publishing Service, Canberra
Hodgkinson J (1979) Farmers' new leader. Canberra Times, 21 July 1979
Howe B, the Hon MP (1989) Wheat Marketing Amendment Bill 1988: Second Reading Speech. House of Representatives Hansard, 19 October 1988
IAC (Australia. Industries Assistance Commission) (1978) Wheat stabilization. Report no 175. Australian Government Publishing Service, Canberra, 30 June 1978
IAC (Australia. Industries Assistance Commission) (1983) The wheat industry. Report no. 329. Australian Government Publishing Service, Canberra, 29 September 1983
IAC (Australia. Industries Assistance Commission) (1988) The wheat industry. Report no 411. Australian Government Publishing Service, Canberra, 25 February 1988
IC (Australia. Industry Commission) (1991) Statutory marketing arrangements for primary products. Report no. 10. Australian Government Publishing Service, Canberra, 26 March 1991
Kelly P (1994) The end of certainty: power, politics and business in Australia, Revised edn. Allen & Unwin, St. Leonards
Kerin J, the Hon MP (1984) Wheat Marketing Bill 1984: Second Reading Speech. House of Representatives Hansard, 13 September 1984
Lloyd AG (1970) Some current policy issues. Aust J Agric Econ 14(2):93–106
Lloyd AG (1982) Agricultural price policy. In: Williams DB (ed) Agriculture in the Australian economy, Second edn. Sydney University Press, Sydney
Makeham JP, Bird JG (eds) (1969) Problems of change in Australian agriculture. University of New England, Armidale
Mauldon RG (1968) Introducing the small farm problem. Farm Policy 8(3):63–67
McKay DH (1965) Stabilization in agriculture: a review of objectives. Aust J Agric Econ 9(1):33–52
McKay DH (1967) The small-farm problem in Australia. Aust J Agric Econ 11(2):115–132
National Farmers Federation (1981) Farm focus: the 80's. NFF, Canberra
Rattigan A (1986) Industry assistance: the inside story. Melbourne University Press, Melbourne
Royal Commission on the Wheat Flour and Bread Industries (1935) Second report. Government Printer, Canberra
Thacher D, Rein M (2004) Managing value conflict in public policy. Governance 17(4):457–486
The Land (1979) Performance counts, not numbers, 26 July 1979
Thelen K (1999) Historical institutionalism in comparative politics. Annu Rev Polit Sci 2:369–404
Warhurst J (1982) The Industries Assistance Commission and the Making of Primary Industry Policy. Aust J Publ Admin XLI(1):15–32
Watson AS, Parish RM (1982) Marketing agricultural products. In: Williams DB (ed) Agriculture in the Australian economy, Second edn. Sydney University Press, Sydney
Whitlam EG, the Hon (1973) Industries Assistance Commission Bill: Second Reading Speech. House of Representatives Hansard, 27 September 1973
Whitwell G, Sydenham D (1991) A shared harvest: the Australian wheat industry, 1939–1989. Macmillan Australia, South Melbourne

Chapter 5
From Domestic Deregulation to Privatisation

Keywords Australian Wheat Board • Deregulation • Privatisation • Institutional change • Rural policy development

This chapter considers the first major exogenous shock to collective wheat marketing in Australia. Following consideration of the Industries Assistance Commission's 1988 report discussed in the previous chapter, the Australian Government passed the *Wheat Marketing Act 1989*. This legislation finally implemented the IAC's recommendations and deregulated the domestic market for wheat, although it left the export monopoly intact. The externally-generated challenge to the Wheat Board's powers triggered a response in the institution which allowed it to protect its core values while adapting to its changed environment. The Grains Council of Australia, by 1989 heavily invested in the preservation of collective marketing, and consequentially the Wheat Board's activities, protested against but was unable to prevent the policy change. Its failure to avoid deregulation of the domestic market caused the GCA to rethink its approach to policy and to initiate a process of strategic planning which ultimately led to even greater institutional change – the privatisation of the Australian Wheat Board.

An important part of the historical institutionalist analysis is the concept of 'critical junctures'. Much of the writing in this field, particularly the early scholarship, emphasised the path dependence of institutions and their staying power. This opened the approach to the reasonable criticism that it was limited in its capacity to explain change. The concept of the critical juncture attempted to address this criticism by emphasising the importance of exogenous influences, or shocks, on the fortunes of an institution. This remained a crude and not entirely satisfactory response; partly because it denied the role of agents internal to the organisation and their capacity to influence organisational evolution from within, and partly because it did not recognise the transformative potential of incremental change. More recent work by Streeck and Thelen (2005a) and their colleagues addresses the issue of incremental change. In their introduction to their collection of case studies, the

L.C. Botterill, *Wheat Marketing in Transition: The Transformation of the Australian Wheat Board*, Environment & Policy 53, DOI 10.1007/978-94-007-2804-2_5,

editors propose five 'broad modes of gradual but nevertheless transformative change' that occur within institutions; namely displacement, layering, drift, conversion and exhaustion (Streeck and Thelen 2005b: 19). They point out that there 'are severe limits to models of change that draw a sharp line between institutional stability and institutional change and see all major changes as exogenously generated' (Streeck and Thelen 2005b: 8). This brings a much more realistic interpretation to institutional transformation and emphasises that quite fundamental change can occur apparently smoothly and with minimal disruption. Of the five forms of institutional change identified, institutional layering provides the closest description of the changes generated by the 1989 domestic deregulation. Layering introduces new ways of behaving that can be sold as consistent with the goals of the institution but set in train changes which can fundamentally alter the institution's evolution. Pierson refers to the 'the creation of "parallel" or potentially "subversive" institutional tracks' (Pierson 2004: 137). The introduction of competition in the domestic wheat market generated this type of transformation as new elements within the Australian Wheat Board responded to incentives that were also new to the organisation. The change to the Wheat Board's legislative framework and the removal of its domestic monopoly was a significant change; to many in the industry it was certainly a 'shock'. Nevertheless the response by the institution can more effectively be understood in terms of the strategies that Streeck and Thelen describe than as a critical juncture.

The Australian Wheat Board continued to survive post 1989 and continued to embody, at least rhetorically, the collectivist values on which it was founded. The changes generated within the Wheat Board highlight an important point about institutions; they can publically proclaim one set of values in order to bolster their legitimacy in the eyes of their supporters while internally pursuing a quite different set of values. The values focus of historical institutionalism is one of its strengths but an important distinction needs to be made between rhetorical values and the values that direct and drive institutional action. The Board's rhetorical commitment to collective values was important in sustaining the support of the Grains Council of Australia, which largely continued to subscribe to these values. The 1989 decision exposed the degree to which the grower body, the GCA, remained embedded in the original values of the Wheat Board. It vehemently resisted change to the point of dealing itself out of discussions with the government. In the process it learnt the lesson that it needed to be more pro-active about the future of wheat marketing and could not assume that the *status quo* would continue indefinitely.

The discussion in this chapter is structured as follows. It begins with a description of the changes to the collective marketing arrangements that occurred with the passage of the 1989 legislation and the responses of the Australian Wheat Board and the Grains Council to the new legislative environment. Important core values at the heart of the marketing arrangements were retained as the export monopoly survived the reform, but the institutional response to the end of the domestic monopoly sowed the seeds for change which were ultimately at odds with these values.

It then describes the next, and arguably most dramatic phase, in the Wheat Board's evolution – its rather peculiar privatisation. This development was the result of a process initiated by the Grains Council, but the outcome was a hybrid institution that contained a number of internal contradictions that eventually could only be resolved by the end of collective wheat marketing in Australia.

Institutional Change and Adaptation

Against the background of economic deregulation described in the previous chapter and following receipt of the Industries Assistance Commission's 1988 report, the Australian Government decided to end the Australian Wheat Board's monopoly in the domestic market. In introducing the legislation into Parliament, the Minister for Primary Industries and Energy argued that there was 'no justification for continuing the AWB's domestic marketing monopoly' (Kerin 1989). The Australian Wheat Board's compulsory powers of acquisition were removed and arrangements for administered domestic wheat prices were terminated. The Guaranteed Minimum Price was replaced by government underwriting of such Wheat Board borrowings on the commercial financial market that were necessary to fund the first advance payment. The legislation also introduced a levy-based Wheat Industry Fund to provide the Australian Wheat Board with a capital base to expand its commercial activities. The specific uses for the Fund were to be determined by the Board in consultation with the Grains Council of Australia. The government guarantee on Board borrowings was intended to be a temporary measure as the legislation made it clear that it would be progressively reduced and be subject to review within 5 years (Kerin 1989).

The Grains Council opposed deregulation and attracted criticism from the Minister for its position. The Minister told Parliament that

> It is a matter of some regret that the GCA has been unable to articulate a vision for the future of its industry and has adopted a position of non-negotiability on the major issues.
>
> While the GCA has been arguing for the status quo it seems not to understand that this is an industry which has been undergoing continuous and, in some cases, quite major change. (Kerin 1989)

Whitwell and Sydenham (1991: 223) describe the Grains Council's position as 'a bulwark against change'. In spite of their close association with the policy making process, the GCA's claim to be the representative of Australian wheat growers was tenuous. While the Council remained a staunch supporter of the Australian Wheat Board's institutional arrangements and its central role in wheat marketing, there was impetus for change coming from individual growers outside the farm organisations. This was emerging particularly among growers in Queensland and northern New South Wales where farmers who were often already operating in deregulated markets for cotton and sorghum, could see the advantages of competition and had the confidence to take on the risk of selling their crop into a contested market.

For the Australian Wheat Board itself, the 1989 legislation triggered major cultural change within the organisation. The organisation needed to adapt from its pre-deregulation position of being a 'marketer as distinct from a trader' (Ryan 1984: 124). Once it was competing on the domestic market, the Australian Wheat Board needed to employ traders. It needed skills in risk management to take positions in the domestic market in order to compete effectively. Traders faced a different set of incentives from the marketers. Insiders in the Australian Wheat Board at the time have suggested that there was a tension within the organisation between the old-style marketers involved with pool operations and the newly employed traders. The latter were more driven by their personal ambition to succeed within the organisation than the old collective notion of delivering outcomes for the growers. An additional issue, which became the subject of subsequent debate surrounding the operations of the Australian Wheat Board, was the lack of effective structural separation between the export pool operations and the cash sales of the Board's Trading Division. There was no effective 'ring fencing' of the pool, raising concerns about cross-subsidisation between the pool and the Australian Wheat Board's activities on the domestic cash market following deregulation.

The Grains Council of Australia was, like its predecessor the AWF, firmly embedded in the wheat marketing arrangements and this continued after 1989. The legislation prescribed at least one consultative meeting per year between the Board and the GCA; and the Minister indicated that he 'would expect the AWB and GCA to consult more frequently than this' (Kerin 1989). An important element of this arrangement was the provision in the Act that 'arrangements entered into by the Board in relation to a consultation may include the Board's agreeing to meet expenses reasonably incurred in relation to the consultation by the Grains Council' (s10(3)). By the 1990s these expenses involved the costs of transporting members of the Grains Council's Executive from across Australia and the staff of the Secretariat in Canberra to Melbourne and hosting a dinner with members of the Board. The GCA often piggy-backed Council Executive meetings off these consultative meetings so that it was not meeting the travel costs associated with its own Executive Meetings. The Wheat Board also on occasion funded international travel to conferences and other events by grains industry representatives outside the formal consultation requirements.

The Minister had anticipated that, following deregulation, the Australian Wheat Board would 'remain a force in the domestic market' and would in fact be 'significantly strengthened' by the 1989 legislation (Kerin 1989). Industry insiders report that other potentially large traders were initially reluctant to enter the domestic market due to pressure from the Wheat Board and for a period there were no big competitors on the market. Competition began to emerge from growers themselves and country-based grain merchants. It was nearly 5 years before a big player, Graincorp, entered the market to compete with the Australian Wheat Board.

The deregulation of the domestic wheat market had repercussions at the political level as well; specifically for the Opposition coalition parties. A long-time supporter of the Australian Wheat Board, the country-based National Party vigorously opposed the deregulation of the domestic market. Its Liberal coalition partners supported the Government's legislation, a pattern to be repeated in 2008 when another Labor government finally moved to abolish the wheat export monopoly. While the Opposition moved a number of amendments to the 1989 legislation on which both parties agreed, the tone of their interventions in the Parliamentary debate was notably different. National Party members such as Bruce Lloyd spoke of their 'anger' at the legislation and predicted 'chaotic months ahead' (Lloyd 1989). By contrast, Liberal Party members highlighted the 'tremendous opportunities' offered by the change and the 'compelling reasons for the deregulation of the domestic wheat market' (Hawker 1989). When the Bill was put to the vote in the House of Representatives, it passed with 116 votes in favour and only 4 against; all members of the National Party of Australia but not all of the National Party's Parliamentary representatives.

The deregulation of the domestic market in 1989 was not surprising in the deregulatory environment which prevailed at the time. What is of interest is the Australian Wheat Board's response and the consequences of that response for the organisation's operations. The changes of the 1980s removed important elements of the system of 'orderly marketing' that had been set up in 1948. No longer was the Australian Wheat Board the central part of a system of collective marketing which took care of a wheatgrower's crop, spread the costs and the risk, and delivered a government-guaranteed minimum price for an undifferentiated commodity. Crop segmentation and differential returns for different grades of wheat, and opportunities to sell grain directly to buyers and through the permit system introduced growers to opportunities that they were already experiencing with other crops such as sorghum and cotton. Pressure for the removal of the domestic monopoly came from sectors of the grower community that disagreed with the peak industry body's adherence to orderly marketing principles. Once the domestic market was deregulated, the Australian Wheat Board as an institution underwent an important culture change. The idea that an institution might adapt its norms and values in response to the types of individuals it recruits is not surprising (see for example Peters 1999: 37). From a slightly paternalistic position as marketer of the wheat crop with a mandate to maximise returns to growers, the Australian Wheat Board now needed to employ traders who could operate on the domestic wheat market in competition with other grain traders using sophisticated marketing techniques. The reward structures for this new breed of Australian Wheat Board staff encouraged a degree of risk taking and their career prospects were related to performance. The move into trading created a clash of cultures within the organisation between the marketers and the traders. It also created problems for farmers who were enticed into high risk hedging positions by traders. The step from the protection of delivering the crop to a pool to exposure to sophisticated financial instruments left many farmers with financial losses, particularly when their crops failed. The Wheat Board and its supporters in the Grains Council had managed to resist the Industries Assistance Commission's calls for complete deregulation and protect single desk marketing but it was clear that the pressure for change was not over.

Some Context: The National Competition Policy

Before progressing the story of the Wheat Board's institutional evolution, it is important to explain a critical economic policy development that occurred in the early 1990s in Australia and which formed the backdrop to the privatisation debate. In 1992 the Prime Minister, Paul Keating, commissioned an inquiry into a National Competition Policy chaired by Professor Fred Hilmer (Hilmer et al. 1993). The National Competition Policy was part of the ongoing microeconomic reform agenda that had been pursued by successive Labor governments since the early 1980s. The report made recommendations to government in six key areas:

- extension of the reach of the Trade Practices Act 1974 (TPA) to unincorporated businesses and State and Territory government businesses;
- extension of prices surveillance to State and Territory government businesses to deal with those circumstances where all other competition policy reforms had proven inadequate;
- application of competitive neutrality principles so that government businesses do not enjoy a competitive advantage simply as a result of public sector ownership;
- restructuring of public sector monopoly businesses;
- reviewing all legislation which restricts competition; and
- providing for third party access to nationally significant infrastructure. (National Competition Council 1998: 4–5)

There was clear potential for the grains industry to come under scrutiny on a number of these points, particularly with regard to public sector monopoly businesses and in the process of legislative review of anti-competitive legislation. The Inquiry report specifically identified anti-competitive practices within agricultural marketing including 'compulsory acquisition' and 'monopoly marketing arrangements' (Hilmer et al. 1993: 141). They argued that practices of this nature were 'often grossly inefficient' (141). The Inquiry argued that there was a need for a 'systematic review of regulations that restrict competition' and that '*there should be no regulatory restriction on competition unless clearly demonstrated to be in the public interest*' (1993: 190 – italics in the original).

The Hilmer Report was followed in 1995 by the negotiation of a series of Competition Policy Agreements between the Commonwealth and State governments which comprised a National Agenda for Microeconomic Reform, a Competition Principles Agreement, a Conduct Code Agreement and an Agreement to Implement the National Competition Policy and Related Reforms (National Competition Council 1998). Part of the process of implementation of the Agreements was the establishment of a process of legislative review with the goal of eliminating anti-competitive elements from both Commonwealth and State legislation and to give effect to the principle of competitive neutrality. The release of the Hilmer Report in 1993 added a further impetus to the grains industry's consideration of its future structures as the prospect of a review of the wheat marketing legislation focused the minds of grains industry leaders. Although the single desk marketing arrangements were an article of faith to many in the industry, there was concern that a review of the monopoly against the criterion of net public benefit (as against net grower benefit) could see the arrangement under serious threat.

The Privatisation Process

Following the 1989 domestic deregulation debate, the Grains Council of Australia initiated the *Grains 2000* project. The project was driven by a perception among industry leaders that if the grains industry did not take control of its own future, it would be subject to potentially undesirable change. This pro-active approach contrasted with the industry's strong stance in the late 1980s when the GCA 'steadfastly refused to accept the possibility of the domestic market being deregulated' (Whitwell and Sydenham 1991: 225). This intransigence had meant that growers had little influence over the structure of the reformed domestic market. Industry leaders in the early 1990s learnt this lesson and were determined to take control of their own future.

Preliminary work was undertaken by consultants, funded through the Grains Research and Development Corporation, to analyse the industry 'from paddock to plate'. An industry conference was held in September 1991 to discuss the way forward. The then Executive Director of the GCA later wrote 'The critical aspect of the Grains 2000 project was the very vivid expression of the industry taking the initiative in strategically planning its direction' (Hooke 1993: 39).

The First Attempt at Change: The Newco Debate

Following the *Grains 2000* Conference, the industry established a Working Group comprising the Grains Council of Australia, the Australian Wheat Board and the Commonwealth Department of Primary Industries and Energy. In April 1992, the Working Group issued a booklet entitled 'A Progress Report on a successor to the Australian Wheat Board'. The report proposed 'a structural and operational framework for a successor to the Australian Wheat Board through the establishment of a private marketing corporation owned and controlled by grain growers' (Grains Council of Australia 1992: 3). The proposal was for the establishment under special legislation of a private company, named 'Newco' in the proposal, which would be owned and controlled by grain growers. The levy-based Wheat Industry Fund would cease to accumulate and equity in the fund would be converted to shares in the new entity. Under the proposal, the Government would legislate to transfer the export monopoly to the new company and, during a nominated capital 'build up' phase, would continue its financial support in the form of the government guarantee on borrowings for the first advance payment. The Government would also grant the body tax-exempt status.

Although the Department of Primary Industries and Energy was listed as a co-author of the proposal, the booklet was clear that 'no indication has been given by the Commonwealth Government as to their thoughts regarding such specific tax rules' (Grains Council of Australia 1992: 19). The Minister for Primary Industries and Energy later indicated that he had 'reservations about some aspects of the proposal' (Crean 1992). The model proposed that the Government would appoint a single director to the board of Newco in recognition of its continuing role in terms of the granting of the export monopoly and ongoing financial support. Share ownership would be limited to a maximum of 5% of the company for any one shareholder and ownership would be limited to graingrowers, statutory marketing and handling authorities for grains, and grain cooperatives (Grains Council of Australia 1992: 10–13) The report was confident that

> The nation should recognise the benefits to it in the proposal and would make its contribution to Newco through specific tax rules, continuing the current financial support arrangements, at least initially, and continuing the industry's single export desk status for wheat. (Grains Council of Australia 1992: 20)

Following the release of the Progress Report, 22 grower meetings were organised to discuss the proposed restructure. The majority of growers did not support the details of the proposal (Crean 1992) but the process generated a series of principles which growers endorsed:

- Retention of the AWB's single desk wheat export seller status;
- Continuation of the Commonwealth Government's underwriting of AWB borrowings;
- The necessity for the AWB or any successor to have an adequate capital base; and
- A capacity to broaden the industry's horizons to facilitate the marketing of products and continued expansion into downstream processing and other value-adding activities. (Hooke 1993: 44)

The Newco proposal failed for several reasons; the simplest of which was that the time was not right. The idea was being promoted to growers only a few years after large parts of the industry had indicated that they were not in the mood for change and had opposed deregulation of the domestic market. The proposal put to the growers for their consideration was light on detail, there was a perception that the consultation process surrounding the proposal's development had been poor and growers simply did not trust the process. There was also concern among growers that 'part of the hidden agenda of the Newco proposal was to get rid of the single desk selling status' (Crane 1992). This particular point was emphasised by opposition members in Parliament who sought to paint the Newco proposal as a Government scheme to remove the export monopoly (see for example Anderson 1992; Fisher 1992).

The promotion of the Newco concept was complicated by the presence at the grower meetings of advocates of an entirely different model, 'Valco'. Valco was an alternative proposal that sought to combine wheat marketing with storage, handling and transport, creating a vertically integrated company with a natural monopoly. The proposal was developed and promoted by the Queensland Graingrowers Association which sent a representative to all the Newco meetings and managed to persuade the organisers to allow him to speak. Queensland growers were particularly suspicious of the Newco proposal, seeing it as a threat to the newly created Queensland grower cooperative bulk handler Grainco and also as a hastily prepared proposal which closed off consideration of other options. Valco was less a serious proposal for change than a tactic designed to highlight to growers that there was, or should be, more than one option put forward for their consideration. It also reflected a perception among State farming organisations and the bulk handlers that the Wheat Board was seeking a dominant position in the market and would seek to take over the operations of the State-based bodies.

In October 1992 the Commonwealth Government introduced legislation which met some of the objectives of the Working Group which had developed Newco by allowing for a portion of the Wheat Industry Fund to be invested in value-adding activity. The legislation extended from 1994 until 1999 the build up of the Wheat Industry Fund as well as the government guarantee on borrowings, and provided for the Australian Wheat Board to establish subsidiaries to undertake value-adding activities. On the export monopoly, the Minister argued that the time was not right to consider changes to the single desk arrangements. He told the Parliament that 'All honourable members will be aware of the corrupt state of the world wheat market where the subsidy war between the United States and the European Community has been waged for many years now' (Crean 1992). However he also stated that 'The Government will review these export powers after the current GATT round [the Uruguay Round of multilateral trade negotiations] outcome is clearer and as any successful reforms arising from the round become effective' (Crean 1992). The Minister also put the industry on notice of further change announcing that 'At the end of the [Wheat Industry Fund capital] build-up period, the Government's firm intention is that the government guarantee will cease as it will no longer be necessary for the Wheat Board's operations. It is also the Government's view that the Wheat Board will be able to be corporatised at the end of the build-up period' (Crean 1992: 2153).

Although the Newco proposal failed, the industry noted that 'The new legislation embodies not so much a change in our institutional structures as a shift in attitude, both within government and across the industry' (Hooke 1993: 45). The failure of the Newco idea was clearly not the end of the debate over the future of the Australian Wheat Board.

Hastening Slowly: The Second Debate

Following the passage of the 1992 legislative amendments, the industry continued to consider the future of grain marketing through a series of Strategic Planning Units. The Strategic Planning Unit charged with overseeing the process across all sectors of the grains industry was the National Grain Marketing Strategic Planning Unit, the task of which was extended in 1994 to include undertaking and managing the strategic planning process in relation to milling wheat, predominantly the future structure of the Australian Wheat Board (Grains Council of Australia 1994: 15). The Strategic Planning Unit commissioned consultants Booz Allen and Hamilton to undertake a study as input into the committee's deliberations.

At the industry's annual conference, Grains Week, in 1995, the Executive Director of the Grains Council emphasised the need for change arguing that 'The AWB is not sustainable in its present form – and it cannot revert. The WIF [Wheat Industry Fund] is not sustainable in its current form'. He also argued that the single export seller status 'must be reviewed continuously as international trade liberalisation progresses' (Hooke 1995: 11). The message from the Council's President was a little more cautious: 'The industry will need to consider a new, perhaps fully privatised, structure for the AWB, provided of course it maintains the single desk' (Macfarlane 1995: 2). As well as arguing in favour of the retention of the single desk he also flagged the preservation of the other key pillar of the collective model, grower control, stating that 'this may be our last chance to get it right and retain grower control of our organisations, our industry and our destiny' (Macfarlane 1995: 2).

Of central concern to the industry was the future of the export monopoly. It was therefore notable that the consultants reflected the Government's view that any benefit from the export single desk was unlikely to remain once reforms from the Uruguay Round of multilateral trade negotiations took hold. They recommended that the issue of export deregulation be kept under review. In briefing the Minister on the consultants' report, the Department of Primary Industries and Energy was careful to point out that the report was targeted at growers and did not take account of the costs or benefits of the single desk to the broader community, the test that would apply under the National Competition Policy which had been agreed in 1995. Even within the narrow confines of quantifying the value of the export monopoly to growers, the consultants' report was lukewarm – suggesting that the single desk was worth between minus $1.75 to plus $7.10 per tonne to growers. A margin this small could easily disappear with movements in exchange rates or underlying world prices.

Following delivery of the consultants' report on wheat marketing in 1995, the Working Group comprising the Grains Council of Australia, the Australian Wheat Board and the Department of Primary Industries and Energy was reconvened with the task of 'developing a paper on the options for a future structure and operation of the Australian Wheat Board to be considered by the [Grains] Council's State affiliates'. The Grains Council's objectives were clear from the outset: 'We will need to design a structure that preserves the AWB's single desk wheat exporter seller status and ensures grower ownership, yet enables the AWB greater commercial freedom and focus' (Grains Council of Australia 1995d). Within the Working Group, differing positions quickly emerged. The Australian Wheat Board's approach to change had shifted markedly from its position during the debate in the late 1980s. It was no longer opposed to change but became an active advocate of the privatisation option, as long as it retained the single export desk. This response appears to indicate that it recognised that the statutory marketing model was no longer viable in the face of the deregulatory trend in government economic policy. However, it continued to seek to protect its core value in the form of the monopoly. It should be noted at this point that while the Board was able to dress up its support for the single desk in terms of the collective values on which it was founded, the monopoly would also be a very valuable asset in the hands of a private company. The Grains Council was more wary of change and more inclined to consider various corporatisation options. Sceptics within the grains industry attributed the Australian Wheat Board's enthusiasm for privatisation to its keenness to gain control of the growing Wheat Industry Fund.

Between April and September 1995 the Working Party worked on an options paper to form the basis of grower consultations.[1] An abridged version of the options paper was produced entitled *Deciding our future: the structure of the AWB* which was mailed to around 40,000 growers. This paper was sent out as background to a series of grower meetings which were held across the wheat belt in September and October 1995; coincidently the same number, 22, as had been held during the Newco debate. Reflecting concern that the Newco proposal had failed to make the case for change, the 1995 paper included a section entitled 'Why more change and why now?'. This identified the following pressures on industry structures:

- The end to the Government guarantee on borrowings in 1999 or, at best, continuing uncertainty over the future of the government guarantee
- A review by the Government of the single export desk in light of GATT Uruguay Round outcomes
- A review under National Competition Policy by the Commonwealth and the States of all statutory bodies, including the statutory grain marketing authorities
- The certainty of increasing pressures on the AWB's operations. (Grains Council of Australia 1995a)

Five possible options were identified for the future of the Australian Wheat Board: reregulation, total deregulation, the status quo, corporatisation with the single desk and privatisation with the single desk. The paper gave short shrift to the reregulation and deregulation options and explicitly stated that the Grains Council had identified a number of key objectives for the new structure. These were:

- Retention of the AWB's single desk
- An adequate capital base to ensure a strong commercial entity with the ability to maintain adequate first advance payments
- A commercial structure which reflects market signals, provides commercial flexibility and maximises returns to growers
- Grower control and/or ownership of the AWB with the ability for growers to access their equity in the AWB
- Industry self-determination, and certainty and efficiency in structural arrangements. (Grains Council of Australia 1995a)

Stated in these broad terms, these objectives were consistent with the approach of the Australian Wheat Board. The paper reiterated the Grains Council's concern that the industry needed to take control over its own future rather than having it imposed by government, arguing that 'Rather than face the uncertainty of enforced

---

[1] At this time, the author was Manager, Strategic Planning at the GCA and was heavily involved in the drafting of the options paper and the subsequent grower meetings.

change, it is preferable for industry to take the initiative' (Grains Council of Australia 1995a). Again industry preference was expressed, as the paper argued that

> Taking the initiative now on the challenges facing the AWB is the best way to promote those things we regard as most important, such as the single desk and grower control of the AWB. (Grains Council of Australia 1995a)

The grower meetings in September and October 1995 each lasted for 3 hours with an introduction by a representative of the local grains industry followed by presentations from Ian Macfarlane, President of the Grains Council of Australia, and Trevor Flugge, Chairman of the Australian Wheat Board. It was agreed in advance of the meetings that, while the Wheat Board would present its preferred position, the Grains Council would outline the various options but argue that the decision rested with the growers. A change consultant was employed to advise the Grains Council on convincing growers of the need for change and she provided input into the drafting of Macfarlane's speeches and his mode of presentation. The presentations by Flugge and Macfarlane were followed by a lengthy question and answer session from the floor and these discussions became quite heated at times, particularly as individual growers questioned the need for change. The Australian Wheat Board's preferred position was for the Board to be privatised but to retain its export monopoly; essentially the Newco model although all participants were very careful not to mention Newco during the debate.

The 22 meetings were taped and the resulting 66 h of tapes were then analysed by the Grains Council to identify and distil the key issues of concern to the growers present at the meetings. In addition, a brief questionnaire was also prepared which was circulated at the meetings and was also available to growers who did not attend the meetings. Consideration of 850 grower survey forms that were completed and analysis of the issues raised at the meetings resulted in the allocation of priorities between the objectives identified by the GCA in its discussion paper. The retention of the Wheat Board's single desk was identified as the top priority with grower control and/or ownership next. Growers added their own objectives to the survey forms and the most common additions related to the composition of the Board, the importance of preventing non-grower control of the restructured Wheat Board and the avoidance of foreign involvement. The analysis of the questions raised from the floor during the meetings reinforced the focus of grower concern on the single desk with 113 questions out of a total 390 relating to its retention. Of these, 23 expressed concern that a privatised body might not retain the export monopoly, an issue that had been raised by opposition members in the Parliamentary debate over the 1992 legislation following the failure of the Newco proposal. The next greatest concern was grower control (69 questions) with a third of these relating to concerns about the tradability of shares and the potential for takeover by non-growers (Grains Council of Australia 1995c). Reflecting the uneasiness that persisted with the clash of cultures within the Australian Wheat Board that had followed domestic deregulation in 1989, a small number of growers expressed concern over the separation of the Wheat Board's cash trading operations from the export pools.

After the meetings, the GCA Executive agreed that the strategy of approaching the meetings without a preferred position had been successful and that GCA had met its objectives from the grower meetings of:

- Acceptance by growers of the possibility of change
- Acceptance of GCA as the growers' representative at a national level
- Identification and prioritisation of the objectives [of the AWB restructure]. (Grains Council of Australia 1995c)

Although the second objective seems unusual, it reflected some internal tensions within the Working Group which had seen the Wheat Board's Chairman Trevor Flugge argue in a telephone conversation that 'the GCA only represents 40% of growers and the AWB represent the lot…' (Grains Council of Australia 1995b). The lack of a clear GCA position was also seen as reinforcing the role of the GCA as the growers' representatives 'in contrast to the AWB's position of a preferred option which had been developed independent of the wishes of growers' (Grains Council of Australia 1995c).

Following the round of grower meetings the differences within the Working Party became clearer. The GCA was concerned that the Wheat Board was understating the potential impact of National Competition Policy on the restructure process. The concern was that a privatised company with a statutory monopoly would not survive a National Competition Policy legislative review. Macfarlane told the October 1995 meeting of the GCA Executive that 'the Minister, the Shadow Minister and DPIE [Department of Primary Industries and Energy] all question whether the preferred option of the AWB would satisfy the Competition Council' (Grains Council of Australia 1995c). This concern was reflected in a February 1996 draft of an Issues Paper on the Wheat Board restructure in which it was noted that

> Mr John Anderson, Shadow Minister for Primary Industries and Energy, had indicated that a private company listed on the share market may not be granted the single export desk power by a Coalition government. (Grains Council of Australia 1995f: 10)

In late 1995, while the Wheat Board continued to refine a two stage privatisation process, eventually known as the 'transitional model', the Grains Council began work on a novel corporatisation model which gave growers 51% control of the restructured body but left the Government with a particular role in relation to the management of the single desk. In December 1995 the GCA told the Wheat Board that the model it saw as the most achievable involved:

- Grower control of the entity, possibly 51% control
- Corporatised body with a majority elected board
- Some Board members to be selected through a selection process involving the Minister for Primary Industries and Energy
- A special government member to protect the government's interests in relation to the export monopoly
- A grower chairman
- Review of the single desk in 2003
- The cessation of the government guarantee and the Wheat Industry Fund in mid 1997
- Conversion of WIF equity into shares in a corporatised AWB (Grains Council of Australia 1995e).

The '51% Option' as it became known was referred to the Department of Primary Industries and Energy for comment and, following advice from the Attorney-General's Department, rejected as unworkable. The Principal Solicitor for the Australian Government Solicitor was concerned that 'If the company was mismanaged, the Commonwealth may face criticism in relation to its involvement in the company even if its actual liability as a shareholder was limited' (Grains Council of Australia 1996c). Having expressed concern that the 51% option did not protect the single desk arrangement adequately, the Wheat Board continued to pursue its preferred option within the Working Group.

In February 1996 the Grains Council still remained unconvinced by the Australian Wheat Board's model. An Executive teleconference in February decided 'That the GCA not endorse the transitional model as an appropriate model for restructuring the AWB at this stage' (Grains Council of Australia 1996b). This meeting also agreed its election strategy for the March 1996 Federal Election which was to gain commitments from both sides of politics regarding the future of the single desk. This was achieved. Both the Australian Labor Party and the Liberal-National Party Coalition agreed that any National Competition Policy review of the *Wheat Marketing Act 1989* (as amended to implement any restructure) would not take place until the end of the period allocated for the National Competition Policy legislation review, anticipated to be late 2000.

The Grains Council Annual Conference, Grains Week, in 1996 was intended to be decision time for the industry; and there was a decision of sorts. The Grains Council agreed to progress the restructuring of the Australian Wheat Board but it was not yet ready to accept the Board's privatisation model. The motion passed at Grains Week hinted at the privatised model that was later adopted, splitting the Wheat Board's activities into a statutory authority with control of the single desk and pooling operations and a subsidiary which would undertake management of investments, non-pool financing, risk management, and eventually the advanced payment system. The motion went on to direct the GCA/Wheat Board/DPIE Working Group to look at other structural options such as converting the subsidiary into a cooperative operating on cooperative principles; a private company operating on cooperative principles; a public listed company using cooperative principles of control; or a government corporation. These options were to be considered against the five objectives of retention of the single desk, an adequate capital base, a commercial structure which maximised returns to growers, grower control and/or ownership and industry self-determination.

Following Grains Week, the Working Group was re-formed and continued to work on various options for the restructure. Over the balance of 1996 and into early 1997, a variety of alternative models was developed and considered. Consultants were engaged to provide advice on the impact of the different models on the Australian Wheat Board's capacity to borrow cheaply and on the likely reaction of the Stock Exchange to various forms of share structure.

Internal correspondence within the Grains Council indicates the level of concern within the GCA about control of the debate. In October 1996, the Executive Director warned the Executive that if they did not reach an agreed model at their 18 October meeting, the future of the Wheat Board would be out of their hands. He went so far as to argue that

> The credibility of the Council is on the line, and the AWB and DPIE would like nothing better than the Council not be in a position to provide a unanimous proposal from that meeting. My information is that the AWB and DPIE are anxious to proceed without the Council, as they are not convinced that the Council can provide the necessary leadership and decisiveness required. (Grains Council of Australia 1996a)

At Grains Week 1997, agreement was finally reached to what was known as the 'Grower Corporate Model' although there remained areas of disagreement. The Government agreed to give effect to the grower corporate model and this was done in two stages. The first piece of legislation in 1997 amended the *Wheat Marketing Act 1989* to facilitate the establishment by the statutory Wheat Board of a subsidiary company which would go on to become the grower owned entity. This legislation was examined by the Senate Rural and Regional Affairs and Transport Legislation Committee before its passage. The Committee identified a number of concerns with the legislation. A key issue was the two stage process which set in train a restructuring of the Australian Wheat Board before the final shape of the new body was agreed. The Committee noted that the 'core comment in this vein to the Committee centred on the uncertainty about the exact arrangements for the share structure of the new companies formed by the Bill' (Senate Rural and Regional Affairs and Transport Legislation Committee 1997: 21).

Also of concern was the potential conflict of interest between the requirement to maximise returns to growers through the single desk arrangement and the requirement to maximise returns to shareholders under Australian corporations law (Senate Rural and Regional Affairs and Transport Legislation Committee 1997: 29). This issue had been one of the reasons for the caution shown by growers during the restructure debates in 1995 and was colourfully expressed by Mr. Wilson Tuckey MP during the Bill's Second Reading debate. Mr Tuckey, himself a wheatgrower, claimed that the model proposed was 'totally unworkable'. He argued that 'The problem is that the proposal is to create a company that is run by its customers but which seeks over time to attract non-customer share capital. This creates a massive conflict of interest…' (Tuckey 1997: 9138). He went on to point out that

> …the question arises of investors being interested in shares when the directors' main interest is in returns to their customers. Is the AMP going to go and buy a lot of my farmers' shares…? Are they going to buy their shares and say, 'This is a great investment'. The directors do not want to make a profit, They want to return all the money to their customers, the wheat producers.' (Tuckey 1997: 9139)

In spite of his reservations, Mr. Tuckey did not vote against his Government's legislation but he continued to be a vocal critic of the arrangements.

The first tranche of the legislation in 1997 allowed the Australian Wheat Board to establish subsidiary companies to take over some of its functions. The Board itself was left with the statutory functions of 'management of the export monopoly, management of the wheat industry fund, and overseeing the activities of its subsidiary, the holding company'. The legislation suggested that from 1 July 1999 these roles would be taken over by an 'independent regulatory mechanism ... established to manage and monitor the performance of the export monopoly'. The holding company subsidiary was set up under Corporations law with the role of undertaking the former Board's commercial activities – for example, trading in wheat and other grains and managing value adding activities – while a further subsidiary 'Company B' would operate the export pool (Scott 1997: 6475).

The 1998 legislation, which came into effect on 1 July 1999, completed the transition of the Australian Wheat Board to AWB Limited. 'Company B' was privatised through the conversion of equity in the Wheat Industry Fund into shares in AWB Limited. Two classes of shares were established in order to meet the requirement that the new entity be grower-controlled. Each wheatgrower, as defined in the company's constitution, received one Class A share. These shareholders elected the majority of the members of the board of AWB Limited but received no dividends. WIF equity was converted into Class B shares allocated proportionally on the basis of WIF holdings. Class B shareholders elected the minority of AWB Limited board members and had a right to receive dividends. Class B shares became fully tradeable on the Australian Stock Exchange when AWB Limited listed on 22 August 2001. A fully owned subsidiary of AWB Limited, AWB International, was made responsible for running the export pools. For those familiar with the history of the Canadian wheat industry, the two class share model will look very familiar – it is very similar to the structure introduced by United Grain Growers Limited when it reorganised its capital structure in 1941, distinguishing between two classes of shareholders which left the control and direction of the company in the hands of its grower members (MacGibbon 1952: 190).

With the privatisation, the statutory remnants of the old Australian Wheat Board became the Wheat Export Authority. The legislation set out the functions of the WEA as:

(a) To control the export of wheat from Australia;
(b) To monitor nominated company B's performance in relation to the export of wheat and report on the benefits to growers that result from that performance. (*Wheat Marketing Act 1989* s5(1)).

In fulfilling the first of these functions, the Wheat Export Authority issued permits for the export of wheat, which was otherwise a prohibited export from Australia. However, the export monopoly was effectively retained by AWB Limited. The company was exempted from the requirement to obtain a permit for export and the legislation required that the WEA receive written approval from AWB Limited before an export permit could be granted to any other applicant. This constituted an effective veto and from 1999 until 2006, AWB Limited did not agree to the issuing of any permits for bulk wheat exports under this mechanism except for one, which it advised the WEA was an 'error' (Australia. Senate Rural and Regional Affairs and Transport Legislation Committee 2006: 65). The powers of the WEA were much weaker than would be expected of the 'independent regulatory mechanism' foreshadowed in the 1997 legislation. The second reading speech for the 1998 legislation referred to the body as an 'independent statutory authority' with no reference to a regulatory function. The legislation set it up to 'manage' the wheat export monopoly and also to 'oversight the pool subsidiary's [i.e. AWBI's] use of the export monopoly to ensure it is being used in accordance with the intentions of parliament' (Anderson 1998: 3332).

Government Policy and the Privatisation Process

The most striking feature of the process just described was the limited role of Commonwealth Government agencies in the debate over the future structure of the Australian Wheat Board. While a member of the nominally tripartite Working Group, the Department of Primary Industries and Energy took something of a back seat, fielding technical questions about topics such as the Government's interpretation of the impact of the Hilmer report on National Competition Policy on any new structure and on various options relating to corporatisation. An insider in the process reports that the Department was ideologically inclined toward privatisation however they did not push this position actively within the Working Group. The majority of the interaction over the contents of the options paper was between the Wheat Board and the Grains Council. This low-key approach to the Government's role in the process appears to have had ministerial endorsement. At the Grains Week Conference in 1995 the Minister for Primary Industries and Energy, Senator Bob Collins referred to the restructure debate and the issues to be raised and, in concluding his speech, said

> I'm impressed with the industry's commitment to confront the future.
>
> I look forward to seeing the proposals from industry on future grains marketing arrangements. (Collins 1995)

In the body of the speech he emphasised that the grains industry would not be exempt from the impact of national competition policy and also pointed out that 'the question of the single export desk is also an integral part of considering the future shape of the AWB' (Collins 1995).

The 'hands off' approach to the restructure that had been taken by the outgoing Labor government was continued under the new Coalition government after March 1996. Although he has since described the privatisation debate as a 'dog's breakfast' (personal communication, 2007), the new Minister for Primary Industries and Energy and Deputy Leader of the National Party, John Anderson told Grains Week 1996 that 'You should determine your own agenda for change and embrace a program that suits you – rather than let others set the pace through adverse market circumstances or policy changes, in Australia or internationally' (Anderson 1996). The Minister identified 'three fundamental principles' he believed should guide the restructure debate: self-reliance; grower ownership and control of marketing; and a fully commercial approach to marketing. He reiterated the point that he was 'looking to the leadership of the industry to continue to make the decision on how you will deliver these objectives' (Anderson 1996).

There is little available evidence to suggest that either the Department of Primary Industries and Energy or the Department of Finance, which was normally involved in privatisation processes, provided advice to the Working Group about the extent to which any restructuring proposals were consistent with government guidelines or practices relating to the privatisation of public enterprises. The Government, both at Ministerial level and through its public service, was content to allow the industry to set the agenda and plot the course of the privatisation of the Wheat Board. It seems that at no stage were issues or principles raised about the transfer of a publically-granted monopoly into private ownership. The industry itself was clearly concerned about shielding the new structure from National Competition Policy scrutiny due to the perception that it might fail the 'public interest' test when subjected to legislative review. There is a number of reasons for this lack of direction from government. First, the Grains Council of Australia had indicated through its strategic planning process that it wanted to take control of its own future. Although growers came to recognise the advantages of domestic deregulation after the fact, the 1989 changes had been seen as top down and imposed without consultation. The industry wanted to ensure that it was not again the recipient of potentially undesirable regulatory change over which it had no control. Secondly, the Australian Wheat Board itself was anxious to reduce the amount of ministerial control over its activities. In its 1995/96 Corporate Plan, the Australian Wheat Board argued for change, noting that

> Particular problems have … been identified with the current AWB/WIF structure due to the unwieldy WIF accountability provisions and the commercial constraints this places on the operation of the AWB… (Australian Wheat Board 1995: 1)

As an institution it was keen to protect its greatest asset – the export monopoly – but there were also strong incentives for removing the organisation from government oversight.

A third reason for the 'hands-off' approach by government relates to the politics of the non-Labor coalition. When wheat pooling arrangements were proposed by the Scullin Labor government in 1930, the leader of the Opposition, John Latham, argued vehemently against the arrangements, particularly objecting to the potential for the development to be 'a move in the direction of governmental monopoly'. He argued that

> A wheat pool board, seen though to be largely administered by persons in the industry itself … cannot be expected to work as efficiently as if the wheat were handled by the institutions acting upon ordinary business principles. … With a monopoly we lose the tremendous advantage of a comparative standard of business efficiency. (Latham 1930: 1427)

However, the export single desk has been an article of faith for the farmer-based National Party, and its predecessors, as evidenced by the Country Party leader Earl Page's support for the 1930 legislation (Page 1930: 2048).

A fourth explanation could relate to a perennial problem in grains industry politics – the essentially divergent interests between the large export-oriented wheat producers and the smaller growers. As late in the debate as April 1997, the West Australian Farmers' Federation member of the Grains Council's Executive was arguing that the allocation of A Class shares should be weighted to give Western

Australian growers greater power within the new company. The Executive Director of the GCA responded that the Council did not want 'to disenfranchise small growers' (Grains Council of Australia 1997). The crux of this issue is whether grains industry policy should be made for graingrowers or the grains industry. Recent estimates suggest that 80% of the national wheat crop is produced by large growers and 'At the other end of the scale, small-scale wheat growers, who are particularly concentrated in New South Wales and Queensland, now only produce four percent of the wheat crop despite comprising one third of all wheat growers' (Staley 2008: 3). The needs of these two groups are quite different with the larger farmers more inclined towards deregulation than the numerically stronger smaller farm operators. A government which took a strong stand one way or the other on this debate ran the risk of alienating an important portion of the wheat industry.

The privatisation debate also illustrates the closed nature of the wheat policy community that was operating at that time in Australia. Unlike the privatisations of other government entities which took place in the 1980s and 1990s, the Wheat Board privatisation attracted little public scrutiny or debate beyond the grains industry. Even the peak farm body, the National Farmers Federation was uninvolved in the discussions.

Conclusion

The 1990s privatisation debate took place against a backdrop of arguably inexorable forces of economic policy liberalisation. The trend towards the removal of government support for agriculture which had commenced in the 1970s accelerated through the 1980s as a series of reports to government from the Industries Assistance Commission and its successors and government policy reviews consistently raised concerns both with the model of statutory marketing authorities and the monopoly powers held by the Australian Wheat Board. As an institution, the Australian Wheat Board had survived nearly half a century embodying a set of collective values which promoted shared risk through pooling activities, and based in a deep seated belief in the importance of monopoly marketing. Successive pieces of legislation had created close links between the Board and the peak industry body which shared its values and the wheat policy community had successfully held at bay critics who had begun to question the *status quo*. The 1989 domestic deregulation occurred in spite of Australian Wheat Board and GCA wishes and the organisation was then forced to adapt. It achieved this by employing a new breed of employees with different skills sets and different career incentives. The export monopoly remained intact as the cornerstone of the Wheat Board's operations but the attitudes and methods of the traders on the domestic market generated a level of internal cultural change.

The privatised entity continued to embody the key GCA objectives of grower control and preservation of the single desk but, in retrospect, the structure appears to have been doomed from the outset. By limiting the debate in the mid-1990s to exclude discussion of an option which ended the single desk, the discussion focused on second order issues. It did not engage with the rationale for regulation in the wheat industry. It was an article of faith for industry leaders that growers would lose if the export monopoly were wound up. The institutional memory of the activities of grain traders in the 1930s was potent and single desk supporters were adamant that growers would lose if middle men were once again introduced into the wheat market. The Wheat Board of course could see the advantages for a private company of holding a monopoly; it much preferred to operate in an uncompetitive environment and there was no incentive for it to open up debate about the logic of preserving the single desk. Where the single desk had been central to the institution's creation in an era of orderly marketing, it became a valuable commercial asset to a privatised entity.

By 1 July 1999, the Wheat Board had achieved its objective of privatisation while retaining the export monopoly. The Grains Council could argue that it had also achieved its key objectives of preserving the single desk and grower control through the dual class structure. Reflecting the collective values of the old marketing body, the rhetoric surrounding the privatisation and subsequent debates over wheat marketing emphasised the centrality of the 'grower's interest' in the operations of the Wheat Board. However, the 'grower' was always narrowly defined. Opponents of the single desk were successfully excluded from the privatisation debate.

The Wheat Board itself was an active player in the privatisation process which in itself is worthy of note. Wheat Board employees clearly saw personal benefit in privatisation as the remuneration for senior officials in a company the size of the Board would be likely to be much higher than the salaries of public servants in a statutory authority. While it advocated privatisation to remove some of the constraints on its operations, it continued to appeal to core values that had underpinned the 1948 legislation. The most influential sectors of the industry remained steadfastly attached to the collective marketing of the export wheat crop and as noted it was in the Wheat Board's interest to preserve the value at the very heart of the arrangements – the export monopoly. The policy process was dominated by a tight policy community which excluded opposing voices.

As noted earlier, in his work on advocacy coalitions, Sabatier (1988) identifies three levels of beliefs that unite such coalitions within a policy subsystem. These are deep core values, near or policy core values and secondary aspects. Advocacy coalitions will negotiate at the outer ring of these belief systems but changes in deep core values are described by Sabatier (1988: 144) as 'akin to religious conversion'. These deep core beliefs are very resistant to change; they are fundamental understandings about the way the world operates. In the wheat marketing debate, the Grains Council and to a lesser extent, and arguably for different reasons, the Wheat Board, hung on to the export monopoly and grower control of wheat marketing arrangements while accepting government withdrawal in other areas such as loss of the government guarantee. These two features of collective wheat marketing have all of the hallmarks of deep core beliefs. At a political level, the National Party also remained a staunch believer in these elements of the institution.

However, within a decade of privatisation both these prizes, grower control and the export monopoly, had been lost. The cause of their demise was the activities of AWB Limited during the Oil-for-Food program; arguably Australia's largest corporate scandal and subject of the next chapter. Although Oil-for-Food provided the catalyst for the death of collective wheat marketing in Australia, the seeds had been sown back in 1989 when the internal values of the Wheat Board shifted in response to domestic regulation.

References

Anderson J, the Hon MP (1992) Wheat Marketing Amendment Bill 1992: Second Reading Debate. House of Representatives Hansard, 5 November 1992

Anderson J, the Hon MP (1996) The Australian grains industry - the way ahead. Speech by Minister for Primary Industries and Energy at the Grainsweek Annual Conference of the Grains Council of Australia, 17 April 1996

Anderson J, the Hon MP (1998) Wheat Marketing Legislation Amendment Bill 1998: Second Reading Speech. House of Representatives Hansard, 14 May 1998

Australia. Senate Rural and Regional Affairs and Transport Legislation Committee (2006) Estimates (Budget estimates). Committee Hansard Commonwealth of Australia, Canberra, 24 May 2006

Australian Wheat Board (1995) AWB Corporate Plan 1995/96

Collins B, Senator the Hon (1995) Address to Grains Council of Australia's Plenary Session. Grains Week 1995

Crane W, Senator (1992) Wheat Marketing Amendment Bill 1992: Second Reading Debate. Senate Hansard, 2 December 1992

Crean S, the Hon MP (1992) Wheat Marketing Amendment Bill 1992: Second Reading Speech. House of Representatives Hansard, 14 October 1992

Fisher PS, MP (1992) Wheat Marketing Amendment Bill 1992: Second Reading Debate. House of Representatives Hansard, 5 November 1992

Grains Council of Australia (1992) Progress report on a successor to the Australian Wheat Board: a report by the Joint Working Party comprising the Grains Council of Australia, Australian Wheat Board and Commonwealth Department of Primary Industries and Energy. Noel Butlin Archives Centre. N158, Box 4 AW12 AWB Restructure Part 1, Canberra

Grains Council of Australia (1994) Grains industry strategic planning: the processes for inventing the future. Continually updated: current revision: 25 March 1994 Unpublished GCA document

Grains Council of Australia (1995a) Deciding our future: the structure of the AWB. Noel Butlin Archive Centre N158, Box 4 AW12 Part 1, Canberra

Grains Council of Australia (1995b) Discussion between Greg Ellis and Trevor Flugge: memo from Greg Ellis to GCA Executive 10 October 1995. Noel Butlin Archives Centre. N158, Box 20 GE2.1 Part 1, Canberra

Grains Council of Australia (1995c) GCA Executive Committee Meeting 24–25 October 1995: Agenda Papers Noel Butlin Archives Centre. N158, Box 20 GE2.1 Part 2

Grains Council of Australia (1995d) GCA/AWB/DPIE Working Group to develop AWB Options Paper. GCA News, 3 April 1995

Grains Council of Australia (1995e) Minutes of GCA/AWB Consultative Group Meeting 21 December 1995. Noel Butlin Archives Centre. N158, Box 23 GE6.1, Canberra

Grains Council of Australia (1995f) The Structure of the AWB: Issues Paper. Noel Butlin Archives Centre. N158, Box 19 GE2.1 February 1996, Canberra
Grains Council of Australia (1996a) Facsimile from Neil Fisher to GCA Executive Members and Grain Executive Officers. Noel Butlin Archives Centre. N158, Box 4 AW12 AWB Restructure Part 4 July 1996–October 1996, Canberra
Grains Council of Australia (1996b) GCA Executive Teleconference 2 February 1996. Noel Butlin Archives Centre. N158, Box 19 GE2.1 February 1996, Canberra
Grains Council of Australia (1996c) Letter from Christine Cheney, Principal Solicitor for the Australian Government Solicitor to Ian Cottingham, DPIE: proposed restructuring of Australian Wheat Board (AWB) 5 January 1996. Noel Butlin Archives Centre. N158, Box 4 AW12 AWB Restructure Part 2, Canberra
Grains Council of Australia (1997) GCA Informal Executive Committee Meeting 16 April 1997: confirmed notes. Noel Butlin Archives Centre. N158, Box 25 GE2.2 January 1997–June 1997, Canberra
Hawker D, MP (1989) Wheat Marketing Bill 1989: Second Reading Debate. House of Representatives Hansard, Canberra, 3 May 1989
Hilmer FG, Rayner MR, Taperell GQ (1993) National Competition Policy. Report by the Independent Committee of Inquiry Australian Government Publishing Service, Canberra
Hooke M (1993) The Australian grains industry: planning for the future. Agr Sci 6(1):39–45
Hooke MH (1995) Distilling the issues. Grains Week 1995 Grains Council of Australia, Brisbane, 3–6 April
Kerin J, the Hon MP (1989) Wheat Marketing Bill 1989: Second Reading Speech. House of Representatives Hansard, 13 April 1989
Latham J, the Hon (1930) Wheat Marketing Bill 1930: Second Reading Debate. Parliamentary Debates, 2 May 1930
Lloyd B, MP (1989) Wheat Marketing Bill 1989: Second Reading Debate. House of Representatives Hansard, 3 May 1989
Macfarlane I (1995) Report by the President of the Grains Council of Australia. Grains Week 1995 Grains Council of Australia, Brisbane, 3–6 April
MacGibbon DA (1952) The Canadian grain trade 1931–1951. University of Toronto Press, Toronto
National Competition Council (1998) Compendium of National Competition Policy Agreements. 2nd edn. Commonwealth of Australia, Canberra
Page E, Dr the Rt Hon Sir, MP (1930) Wheat Marketing Bill 1930: Second Reading Debate. Parliamentary Debates, Canberra, 15 May 1930
Peters BG (1999) Institutional theory in political science: the 'new institutionalism'. Pinter, London
Pierson P (2004) Politics in time: history, institutions, and social analysis. Princeton University Press, Princeton
Ryan TJ (1984) Wheat marketing. Rev Mark Agric Econ 52(1):117–128
Sabatier P (1988) An advocacy coalition framework of policy change and the role of policy-oriented learning therein. Policy Sci 21:129–168
Scott B, the Hon MP (1997) Wheat Marketing Amendment Bill 1997: Second Reading Speech. House of Representatives Hansard, 26 June 1997
Senate Rural and Regional Affairs and Transport Legislation Committee (1997) Wheat Marketing Amendment Bill 1997. Parliament of the Commonwealth of Australia, Canberra
Staley L (2008) Wheat marketing reform. Submission to the Senate Rural and Regional Affairs and Transport Committee Inquiry into the Wheat Export Marketing Bill 2008 and Wheat Export Marketing (Repeal and Consequential Amendments) Bill 2008. Institute of Public Affairs
Streeck W, Thelen K (eds) (2005a) Beyond continuity: institutional change in advanced political economies. Oxford University Press, Oxford

Streeck W, Thelen K (2005b) Introduction: institutional change in advanced political economies. In: Streeck W, Thelen K (eds) Beyond continuity: institutional change in advanced political economies. Oxford University Press, Oxford

Tuckey W, MP (1997) Wheat Marketing Amendment Bill 1997: Second Reading Debate. House of Representatives Hansard, 2 October 1997

Whitwell G, Sydenham D (1991) A shared harvest: the Australian wheat industry, 1939–1989. Macmillan Australia, South Melbourne

Chapter 6
The Monopoly Wheat Exporter and the Dictator

Keywords Iraq • Economic sanctions • Oil-for-Food Program • AWB Limited • Institutional change

At roughly the same time that the grains industry was debating the future of organisational arrangements for a restructured Australian Wheat Board, important developments were taking place at international level which were to prove the ultimate undoing of the single desk. This chapter discusses AWB Limited's involvement in the United Nations Oil-for-Food program and the implications for the wheat industry when these activities came to light following the invasion of Iraq in 2003. It begins with a discussion of the UN sanctions regime and its impact on Australian wheat exports and then describes the policy changes that occurred following revelations of AWB Limited's role in assisting Saddam Hussein's regime to 'game' the Oil-for-Food program. Although the Oil-for-Food scandal created great consternation in Australia, it is not entirely surprising that the newly privatised company behaved as it did. The chapter draws on the economic sanctions literature in suggesting that the company's activities should have been predictable. It also points to the timing issue. The Oil-for-Food program and the implementation of the privatisation occurred in parallel, blurring the lines between the activities of the government statutory authority and the privatised company.

Iraq, Sanctions and Australian Wheat Exports

The UN Sanctions Regime and the Oil-for-Food Program

After Saddam Hussein's regime invaded Kuwait in August 1990, the United Nations Security Council introduced sanctions against Iraq (Security Council Resolution 661). Resolution 661, which was only the third authorisation of sanctions in the

L.C. Botterill, *Wheat Marketing in Transition: The Transformation of the Australian Wheat Board*, Environment & Policy 53, DOI 10.1007/978-94-007-2804-2_6,

Security Council's history (Malone 2006: 61), called on member states to prevent the import of goods from Iraq or Kuwait, to prevent the export of goods to Iraq or Kuwait and to prevent the extension of credit or other financial or economic resources to Iraq or Kuwait. Humanitarian assistance in the form of medical supplies and foodstuffs could be authorised by a Committee of the Security Council (the '661 Committee') which was established to oversee the implementation of the sanctions. The Australian Government responded to Resolution 661 immediately, introducing an embargo on oil imports from Iraq and Kuwait, a ban on the sale of defence-related equipment to Iraq, and, importantly for the Australian Wheat Board, restrictions on financial transactions (Evans 1990). With the ban on credit arrangements involving the Iraqi Government, the Australian Wheat Board was left with outstanding contracts for nearly $400 million worth of grain as well as being owed the bulk of the $US630 million of Iraqi debt to Australian exporters (Shovelan 1990). Within days of the Australian Government's announcement of the sanctions, the industry's peak body, the Grains Council of Australia, sought a meeting with the Prime Minister to argue that the grains industry was being asked to shoulder the cost of Australia's decision to implement the UN's sanctions. Payments were subsequently made to the Board under export finance insurance arrangements but debate over the balance of the debt continued well into the next decade, with a Senate Inquiry being held as late as 2005 into the repayment of the Iraqi wheat debt (Australia. Senate Rural and Regional Affairs and Transport References Committee 2005).

Resolutions relating to the situation in Kuwait were passed by the Security Council regularly throughout September and October 1990 (Resolutions 662, 664, 665, 666, 667, 669, 670, 674 and 677), covering issues such as the treatment of third country nationals and the use of 'human shields', calling for the release of hostages, calling on Iraq to respond to international demands and reminding the international community of its obligations under Resolution 661. In November 1990, the Security Council adopted Resolution 678 which authorised the use of force against Iraq and a coalition led by the United States commenced an air campaign on 15 January 1990 followed by a ground campaign on 24 February 1991. On 2 March 1991, the Security Council adopted Resolution 686 which established a 'formal framework for a permanent cease-fire' (Malone 2006: 68–73).

On 6 April Resolution 687 was adopted. It was a comprehensive Resolution which reaffirmed the sanctions implemented under Resolution 661 and subsequent resolutions but shifted the objective from obtaining Iraq's withdrawal from Kuwait to the issue of Iraqi disarmament. Resolution 687 also eased sanctions slightly by allowing for the sale of foodstuffs to Iraq but only for cash; the ban on credit sales remained in place. In spite of this concession, it became increasingly apparent during 1991 that the Iraqi people were suffering nutritional and other health problems as a result of the sanctions. In August 1991, the Security Council adopted Resolution 706 which set out the parameters of the program that was to become Oil-for-Food. The extent of the sanctions regime was an 'unprecedented exercise of international jurisdiction over the internal affairs of a sovereign state' (Doxey 1996: 37) with the Security Council effectively 'controlling a sovereign state's revenues and directing its expenditures' (Malone 2006: 116). As Malone notes (2006: 116), 'unsurprisingly, it [Iraq] refused to cooperate'. The idea of using Iraqi oil revenues to pay for humanitarian needs was in response to a concern that a country with the oil wealth of Iraq should not be diverting finite international aid resources away from areas of real need, such as the Horn of Africa (Malone 2006: 115). Nevertheless, with Iraq's refusal to comply with Resolution 706, the UN aid program in Iraq continued to rely on donations until 1996. In April 1996, the Security Council passed Resolution 986 which provided a 'rare concession to Baghdad' (Malone 2006: 117) by giving Iraq responsibility for the distribution of goods under the proposed Oil-for-Food program. In response, the Iraqi Government agreed to the Distribution Plan for Iraq and signed a Memorandum of Understanding with the UN. The program came into effect in May 1996.

Although it wished to respond to the humanitarian needs of the Iraqi people, the international community remained concerned that Iraq not gain access to hard currency with which it could purchase military-related supplies. To facilitate the Oil-for-Food program, an escrow account was set up into which the proceeds of Iraqi oil sales were paid. Approved humanitarian imports were then paid for out of this account. Wheat was an approved import and, as the Australian Wheat Board was keen to recommence sales to the important Iraqi market as soon as possible, it took quick advantage of the opportunities offered by the Oil-for-Food program. It was 'the first western grain merchant back into Iraq following the lifting of the UN embargo on oil sales' (Australian Wheat Board 1996). In announcing its return to the market, the Board explained

> We've maintained close contact with Iraq, potentially one of our largest wheat importers in the Middle East. The Iraqis have continued to buy wheat for cash from us and we're anxious to resume normal business with them as soon as possible. (Australian Wheat Board 1996)

Following the implementation of Oil-for-Food, the Security Council continued to pass resolutions relating to Iraq's weapons program and imposed increasingly demanding inspection regimes. Malone notes that by 1997

> any pretence that mere satisfaction of the disarmament obligations imposed by SCR [Security Council Resolution] 687 would induce the United States to allow sanctions to be lifted had disappeared. [...] Only regime change could satisfy Washington [...] It removed from Saddam Hussein any real incentive to comply with the resolutions. (Malone 2006: 157)

Nations on the receiving end of sanctions are more likely than not to engage in activity to bypass them and Iraq was no exception. The main mechanism Iraq used to bypass the sanctions was oil smuggling, amassing $US11.8 billion (Malone 2006: 129–130) during the course of the program. In light of the attention that the corruption of the humanitarian component attracted, it is worth noting that

> by February 2005 it had emerged that both the Clinton and George W Bush administrations had made formal decisions, of which they notified Congress, to allow much of the oil smuggling to continue, despite its prohibition by both Security Council resolutions and prior US law, since it was in the US 'national interest'. (Malone 2006: 131)

Kickbacks associated with the humanitarian component of the program allowed the Iraqi regime to acquire a further $US1.8 billion in hard currency. From mid 1999, the Iraqi Government began to impose charges on suppliers of goods under the humanitarian component of the Program and by late 2000, 'almost no prospective suppliers of goods to Iraq would see their bids approved by the Iraqi ministries without agreeing to pay a kickback' (Meyer and Califano 2006: 112).

In spite of a number of concerns being raised at the UN about possible bypassing of the sanctions from early 2000, hard documentary evidence of the kickback arrangements only came to light after the program ended in 2003 with the invasion of Iraq and the subsequent demise of Saddam Hussein's regime. In response to the revelations, in 2004 the UN Secretary General Kofi Annan set up the *United Nations Independent Inquiry Committee* into the Oil-for-Food program chaired by Paul Volcker. To the embarrassment of many Australians, the Volcker Report identified AWB Limited as the single biggest contributor of kickbacks to the Iraqi Government through the humanitarian program, having paid $US221 million through the mechanisms established by the regime (Volcker et al. 2005: 331). It was the largest supplier of humanitarian goods under the Oil-for-Food Program, selling $US2.3 billion worth of wheat to Iraq and accounting for more than 14% of the illicit payments made under the humanitarian component of the program (Volcker et al. 2005: 262). The revelations in the Volcker Report prompted the Australian Government to set up its own inquiry, the *Inquiry into certain Australian companies in relation to the UN Oil-for-Food Programme* chaired by the Honourable Terence Cole QC (Cole 2006a). In addition to uncovering the internal workings of AWB Limited's involvement in the kickback scheme, the Cole Inquiry uncovered two other schemes which confirmed concerns about the company's culture and the lengths to which its employees were prepared to go to sell wheat to Iraq. The following reflects the findings of both the Volcker and Cole Inquiries.

The schemes set up by the Iraqi Government were aimed at gaining access to the hard currency in the escrow account for purposes other than those specified in Resolution 986. The Resolution allowed Saddam Hussein's government to set the terms of the contracts for the importation of humanitarian goods and this provided the opening for the development of a mechanism for the payment of kickbacks. Under the Agreement with the UN, Iraq undertook to provide the transportation and distribution of humanitarian goods once they arrived in Iraq, except for distribution to Kurdish areas of the country which was undertaken by humanitarian organisations. Neither the agreement between Iraq and the UN, nor Resolution 986 specified how the goods were to be transported beyond the entry points (Volcker et al. 2005: 265), however, the arrangement did not allow for the use of escrow money to cover transport costs once goods arrived in Iraq (Meyer and Califano 2006: 110). In the early phases of the program, Iraq bore the cost of transporting goods from the port of Umm Qasr (Volcker et al. 2005: 266), the delivery point for Australian wheat. However, beginning in June 1999, the Iraqi regime imposed a transportation fee on imports of humanitarian goods under the Oil-for-Food program. These fees inflated the cost of imported goods which were then paid for from the escrow account. Suppliers then paid Iraq, via third companies, in hard currency for the provision of, arguably non-existent, transport services. Except for a few early contracts which explicitly included the transport fees, the payments were disguised in the contracts that were sent to the UN for approval.

There was no doubt that the granting of contracts was contingent on the payment of the fee, which, in the case of AWB Limited was initially $US10/t and escalated to over $US50/t once an 'after sales service' fee was added in 2000. The transportation fee was to be paid to a nominated transport company, generally Alia for General Transportation and Trade (Alia) or Amman Shipping, both based in Jordan and both front companies for the Iraqi Government (Volcker et al. 2005: 270). These companies took a small percentage of the fee and then transferred the rest to Saddam Hussein's government. The exporters included the fees in the contract price for the goods, thereby receiving hard currency from the escrow account, which was then passed on to Iraq through the front companies. The after sales service fee was generally 10% of the value of the contract. Where the original payment had been justified by the Iraqis as covering genuine transportation costs it was incurring in moving goods from port of entry, the new fee was clearly a mechanism for extracting funds from the escrow account.

One of the reasons AWB Limited, and indeed Australian grain growers when they learnt of them, seemed quite sanguine about these arrangements was that the kickbacks did not cost the Australian exporter anything because the fees were recouped from the escrow account. The only time that AWB Limited raised the validity of the fees with the UN was when the Iraqis sought to impose a further 50c/t on the transport fee in 2001 after a contract had been negotiated. This would have not been recoverable by AWB Limited from the escrow account so it would have cost the organisation money (Cole 2006b: xliii). Meyer and Califano note this feature of the scheme: 'Kickbacks became essentially costless for suppliers, and this was a powerful incentive for suppliers to acquiesce in making illegal payments to the Iraq regime' (Meyer and Califano 2006: 112).

The Australian Government's Cole Inquiry concluded that employees within AWB Limited were very aware that the arrangements were likely to be in contravention of the UN sanctions, although there is some doubt that the company's Board was informed of the kickbacks. Some witnesses admitted as much, acknowledging that they knew the fees were going to Iraq (Cole 2006b: xliv). When the transportation fees were first proposed, email traffic within AWB discussed the mechanism by which the fees could be paid. One email joked that options might include 'to use Maritime agents / vessel owners account / or buy a very large suitcase' (Exhibit 0009 AWB.5035.0337). In the end, AWB Limited employed three mechanisms for disguising their payments. First, they persuaded shipowners to pay the fee on their behalf. Some refused to participate in the scheme because they suspected that it breached sanctions while another company had concerns relating to money laundering. The second mechanism was to interpose a third company between AWB Limited and Alia. The Liechtenstein-based subsidiary of UK-based Ronly Holdings was employed for this purpose for a while, for which they were paid a commission of 20c/t. The third mechanism was to omit reference in the short form contract to the payments so they could not be identified by either the Australian Department of Foreign Affairs and Trade or the UN Office of the Iraq Program, both of which had a role in scrutinising the contracts.

The UN did not deal directly with individual companies exporting to Iraq and relied on member states to certify that exports were in compliance with the conditions of the sanctions regime. This meant that contracts were sent to the Office of the Iraq Program for approval through the Australian permanent mission to the UN in New York. Australian companies were required to submit their contracts and a 'Notification or request to ship goods to Iraq' form to the Department of Foreign Affairs and Trade in Canberra which sent it on to its officials in New York. The Cole Inquiry found that there were four ways in which the Department played a role in the Oil-for-Food Program:

- First, the dissemination of general information and the provision of specific advice to exporters or potential exporters about the sanctions and the requirements for participation in the Programme
- Second, the 'processing and transmission to the United Nations of applications for UN approval of contracts for the shipment of goods to Iraq by Australian companies under the Programme and liaison with the Office of the Iraq Programme in relation to such applications
- Third, the granting of permissions to export goods to Iraq under the Customs (Prohibited Exports) Regulations
- Fourth, liaising with the UN 661 Committee (also called the Sanctions Committee) and the Office of the Iraq Programme in relation to alleged or potential breaches of the Programme by Australian companies. (Cole 2006c: 58–59)

As the Cole Inquiry (2006c: 70) acknowledged, the Department of Foreign Affairs and Trade (DFAT) was not in a position to judge whether the price included in the contract had been inflated and if DFAT officers had no reason to suspect that there was a problem with the contracts, their task was basically routine administration. The Inquiry noted that this constituted 'mechanical checking, not approval' of the contracts (Cole 2006c: 70).

A number of factors suggest that it was well understood within AWB Limited that the transportation fees were a mechanism for channelling money to Iraq and not for the provision of a genuine service. It emerged during the Cole Inquiry that Alia's services were enlisted by AWB Limited without the latter conducting due diligence as to the company's background or corporate structure. The Cole Inquiry was told by Alia's general manager, Othman al-Absi, 'I do not recollect ever being asked by anyone from AWB about who owned or controlled Alia, or whether it had connections with the Ministry of Transport. However, Alia's ownership was public knowledge and was not hidden'. Mr al-Absi also told the Inquiry that the company was '49% owned by or on behalf of the Iraqi Ministry of Transport' (Exhibit 141, WST 0007.0001) and that, in relation to wheat shipments by AWB Limited, its role was simply 'to collect inland transportation fees on behalf of the Iraqi State Company for Water Transport [ISCWT]'. He also revealed that 'In the period 1999 to 2003 when Alia was collecting transport fees on behalf of ISCWT, Alia did not provide transportation services … for AWB, or arrange for the provision of any such transportations [sic] services' (Exhibit 141, WST 0007.0001). When AWB Limited needed Alia actually to provide transport services in 2003, it entered into its first trucking contract with the company.

AWB Limited management was alerted to possible problems in the company's dealings in Iraq when it commissioned a report in 2000 by consultants Arthur Andersen to look into 'the existence of illegal or unethical behaviour and any failure of control systems' (Cole 2006b: xxxvii). According to the Cole Inquiry Report, the Arthur Andersen inquiry was initiated by a manager who was concerned that the Ronly Holdings arrangement mentioned above was a mechanism for individual employees within AWB Limited to steal from the company. The Report of the Andersen inquiry identified the trucking fee arrangement as a potential problem for the company, suggesting that 'This type of arrangement could be misinterpreted as a money laundering process' and that 'the issue of trying to use ship owners to make payments on behalf of AWB potentially damaged the reputation of AWB as would the attempt to disguise the transactions' (Cole 2006c: 449). Although a senior manager was tasked with following up on the report's recommendations, nothing was done.

AWB Limited management was once again motivated to undertake an internal review of the company's dealings with Iraq following allegations in 2003 of the payment of kickbacks to the Iraqi regime which named AWB Limited as a participant. 'Project Rose' was initiated by the company to investigate the allegations and report on any possible legal ramifications. Again external consultants were employed to undertake the review. The Board was advised that the investigation found that

1. All AWB contracts were approved by the office of the Iraq Program at the United Nations;
2. No evidence has been identified of any AWB knowledge that money paid to the Jordanian transport firm, Alia, was onpaid to the Iraq regime;
3. No evidence has been identified of payment of funds by AWB to any other person in relation to the OFF [Oil for Food] shipments; and
4. No evidence has been identified of payment of funds to any AWB employee by any other person in relation to OFF shipments (Cole 2006b: lxviii).

In August 2003, legal advice was provided that included the statement that 'It is possible that AWB's conduct has resulted in a contravention by Australia of UN resolution 986' (Cole 2006b: 208). It is worth noting that the results of Project Rose were withheld from the Volcker Inquiry and were among papers over which AWB claimed legal professional privilege during the Cole Inquiry, a claim which was largely denied by the Federal Court of Australia.

As well as using Alia to provide kickbacks to the Iraqi regime, AWB Limited used this mechanism to undertake two other transactions revealed by the Cole Inquiry. The first related to a claim by Iraq in 2002 that several shipments of Australian wheat were contaminated with iron filings, for which the Iraqis sought compensation. Although AWB Limited disputed the facts of the contamination, they were again faced with the challenge of transferring money to the Iraqi Government when such a transfer was in breach of UN sanctions. Coincidentally the shipments in question had been subject to earlier communications between the Iraqi Grains Board and AWB Limited relating to the currency in which the contracts were written. The Iraqis had tried to change the currency after the contract was signed, and after AWB Limited had currency hedging arrangements in place. AWB Limited insisted on the terms of the signed agreement. The value of the claim for the contamination was very close to the difference between the relevant exchange rates. AWB Limited sought advice from the Australian Department of Foreign Affairs and Trade about the appropriate avenues for resolving the compensation claim. The Department's advice was that AWB Limited could either give a discount on a subsequent shipment or pay compensation into the escrow account (Cole 2006d: 222). Neither of these solutions was acceptable to AWB Limited so the transportation fees mechanism was employed. As Commissioner Cole noted, the 'imperative was to retain the Iraqi grain trade, which it feared would be lost if the compensation claim was not paid' (Cole 2006b: lix).

The other transaction which the Cole Inquiry uncovered was more bizarre. It involved another Australian company BHP Petroleum, a donation of wheat which was redefined as a debt and the sale of that debt to a third company, Tigris. This matter was not directly related to sanctions evasion by the Iraqis but was a complicated scheme aimed at facilitating Iraqi repayment of a debt on a shipment of wheat originally donated by BHP Petroleum in early 1996 before the Oil-for-Food program began. BHP Petroleum had made the donation of wheat, purchased from AWB Limited, as a gesture of goodwill towards the Iraqi Government; and with an eye on lucrative oil prospects in Iraq which could be exploited if and when the sanctions were lifted by the UN. When the Oil-for-Food scheme commenced, BHP Petroleum sought to have its donation reclassified as a sale but was firmly rebuffed by the Department of Foreign Affairs and Trade. In 2000, Mr Norman Davidson-Kelly negotiated with BHP to take over its Iraqi interests, including the $5 million 'debt' for the wheat shipment; these interests were assigned to the Gibraltar-based company Tigris. The Iraqi Government accepted the conversion of the donation to a debt and also the transfer of this debt from BHP Petroleum to Tigris. It therefore needed to find a mechanism for paying Tigris. Tigris did a deal with AWB Limited under which the latter would 'load up' the contract price on a shipment of wheat into Iraq and then pass on the money paid out of the escrow account, minus a handsome commission. The contracts surrounding this arrangement were considered by Commissioner Cole to be a 'sham' (Cole 2006b: lxi).

The Problem of Sanctions Implementation

There is a small but growing literature on the ineffectiveness of economic sanctions as a tool of persuasion in international relations. Scholars generally agree that sanctions do not work (Barber 1979; Pape 1997). In spite of this apparent failure the use of sanctions escalated over the last two decades of the twentieth century. The implementation of sanctions raises a number of key issues, several of which are relevant to the AWB Limited case.

First, while economic sanctions are intended to provide leverage to persuade a recalcitrant state to change its behaviour in a particular way, research suggests that they are frequently counterproductive. The first and most obvious way in which sanctions fail is when the target or receiving state rejects the ultimatum it faces. If the target state is not going to comply with international demands, and it seems unlikely that many would (Pape 1997: 93; Renwick 1981: 87), its national effort becomes focused on either bypassing the sanctions through smuggling or other evasion activities, or living with them (Galtung 1967: 393). Doxey (1971: 130) suggests that sanctions can actually generate 'a certain consolidation of public feeling ... [and] a heightened sense of solidarity and national purpose'. The sanctions against Iraq and the requirements placed on Saddam Hussein's regime by the international community have been described as 'the most extreme sanctions in history' (Pape 1997: 106). Under these circumstances it is unsurprising that the Iraqi Government opted to bypass sanctions rather than comply with international demands. Malone notes that 'the Iraq sanctions regime ... demonstrated most clearly how a cunning target government could turn sanctions to its own ends' (Malone 2006: 134). The impact may also be unsuccessful because, in some cases, the target state has actually benefited from sanctions as they have provided a form of tariff protection which has allowed the development of 'infant industries' and been beneficial to the country's balance of payments position. Sanctions against both Italy and Rhodesia had this effect (Galtung 1967; Renwick 1981: 85).

The second problem with sanctions is that they can have an adverse effect on the sending states as well as the recipient (Renwick 1981: 81). There are a couple of key problems for the implementing state. The first is a matter of enforcement. International agreements are commitments by nation states which then need to put in place measures under domestic law to ensure compliance by their nationals with the terms of that agreement. There are several elements to this: ensuring domestic legal structures reflect the international commitment; monitoring domestic companies and individuals adequately to ensure compliance (Barber 1979: 379; Doxey 1996: 101); and preparedness to act against those who breach sanctions (Doxey 1971: 129; Renwick 1981: 78). The Australian Government had inadequate mechanisms in place to ensure that AWB Limited complied with the sanctions. During the Cole Inquiry, critics of the Government argued that it failed to act on reports that the company was breaching the sanctions against Iraq. One of the problems with this argument is that the Government did not have the power under domestic law to compel AWB Limited to provide information about its activities. Changes to Australian legislation in the aftermath of the Cole Inquiry still provide only weak monitoring of sanctions compliance (Botterill and McNaughton 2008). In addition, the international community itself was inconsistent in its enforcement of sanctions. Malone has argued that 'the Sanctions Committee was willing to tolerate some corruption of the program by Hussein, as the price of its very existence'. He goes on to observe that 'Some but too few commentators understood that Member States had been complicit in the corruption of the Program all along...' (Malone 2006: 131)

In addition to ensuring compliance with sanctions, sending governments face the problem that the costs of imposing sanctions are not evenly distributed through the economy, resulting in differing levels of support for action and commitment to implementation across sectors (Doxey 1971: 107; Renwick 1981: 82). As Barber notes, as a result of this inequity, 'feelings of resentment spring up among those who have to pay the highest price, providing a strong motive for evading or modifying the sanctions' (Barber 1979: 377). Iraq was an important market for Australian wheat. At the time the sanctions were introduced it accounted for around 12% of Australia's wheat exports (AWB Limited undated: 5), much of it sold on credit. The Australian wheat industry had been stung by the implementation of the original round of sanctions in 1990 and, although it gained assurances from the Prime Minister that 'the Government would not expect grain growers to shoulder the full burden of the United Nations trade sanctions on Iraq' (Grains Council of Australia 1990), compensation to growers was still under discussion some 15 years after sanctions were announced. The wheat exporter was understandably wary of the later sanctions regime and its potential cost to Australian graingrowers.

The third major problem with sanctions implementation relates to the incorporation of any form of exceptions into the program. Exceptions undermine the likelihood that the sanctions will achieve their goals and increase the possibilities for sanctions-evasion. The Oil-for-Food program was at its heart a complicated exception to the sanctions against Iraq. Meyer and Califano (2006: 1) have described the Oil-for-Food program as 'destined for corruption'. Apart from the fact that the scheme allowed Saddam Hussein to set the terms of the contracts both for oil sales and the importation of humanitarian goods, the UN operation of the scheme was itself corrupt. Meyer and Califano provide an excellent accounting of the problems with the program. In a nutshell, the major flaws in the program were fourfold. First, the process for selecting the bank which housed the escrow account was corrupt. Second, the process for selecting the company to inspect Iraqi oil exports was corrupt. Third, the process for selecting the company which undertook inspections of humanitarian imports into Iraq was corrupt. Finally, and perhaps most importantly in terms of the capacity of the UN to close off any bypassing of the sanctions regime, the Head of the Office of the Iraq Program at the UN was receiving oil allocations from Saddam Hussein via a company in the Virgin Islands. This effectively ensured that any suspicions about the operation of the program were not adequately pursued. These problems all relate to the implementation of the *exceptions* to the sanctions against Iraq which provided opportunities for corruption that would not have existed if the sanctions had been comprehensive and total. The problems associated with including exceptions in the program were compounded by the fact that the list of exceptions grew over the life of the program. As the Volcker Report noted, 'By 2003, the Programme encompassed twenty-four sectors [authorised to sell to Iraq], far beyond the basics of food and medicine ordinarily associated with a humanitarian relief operation' (Volcker et al. 2005: 252). It should be noted however that some good came out of the program, an outcome often lost in the focus on the program's corruption. Malone points out that:

> Over its lifetime, OFF [Oil-for-Food] handled $64 billion worth of Iraqi oil revenues, and served as the main source of sustenance for 6 percent of Iraq's estimated twenty-seven million people, reducing malnutrition among Iraqi children by 50 percent. It underpinned national vaccination campaigns reducing child mortality and eradicating polio throughout Iraq. In addition, it employed more than 2,500 Iraqis. (Malone 2006: 117–118)

Parallel Timelines

An important element of the story of Australian wheat exports and the Oil-for-Food scandal which has been overlooked in much of the commentary (for example Bartos 2006; Overington 2007), is the timing of the kickbacks in relation to the privatisation debate. As outlined in Chap. 5, debate over the future structure of the statutory Australian Wheat Board began in earnest in 1995 with the round of grower meetings in September and October that year. The industry conference, Grains Week, in 1996 did not arrive at agreement over the structure and asked the Working Party to consider various options against a set of objectives identified by the Conference. Iraq agreed in April 1996 to the Distribution Plan which gave effect to Oil-for-Food and the program commenced in May. It was the statutory Australian Wheat Board that proudly announced it was the first Western grain trader back into the country in May 1996 (Australian Wheat Board 1996). There is no evidence that the wheat contracts implemented while the exporter remained a government entity contained any kickbacks. The employees of the Australian Wheat Board became employees of AWB Limited following the passage of the first tranche of the restructuring legislation in 1997. That legislation is quite clear that the company was 'not taken, for the purposes of a law, to be:

(a) a Commonwealth authority; or
(b) established for a public purpose or for a purpose of the Commonwealth; or
(c) a public authority or an agency or instrumentality of the Crown'. (*Wheat Marketing Amendment Act 1997* s35)

The Cole Inquiry revealed that the first contracts containing kickbacks in the form of the transportation fee were signed in the second half of 1999, one in July and two in October. Even though the first contract was signed within days of the formal privatisation, the employees engaged in negotiating the arrangements prior to that time were already in legal terms considered not to be employed by a Commonwealth government agency. It is worth noting that each of these three contracts actually made the transportation fees explicit but this was not picked up by either the UN or the Department of Foreign Affairs and Trade.

The activities of AWB Limited in Iraq became a major political scandal in Australia. The issue dominated the news for much of 2006 and the focus quickly moved to how much the Government knew, who knew and who should have known about the kickbacks. There was discussion about the lack of accountability and speculation that National Party Ministers had chosen to ignore the sanctions busting activities. It is possible that much of the debate around the knowledge of the Australian Government about the kickback arrangements relates to confusion concerning the timing of the wheat exporter's transition to a private company and the point at which its employees ceased to be employed under government legislation. The choice of AWB Limited as the name for the new company compounded the confusion as the old Board had often been referred to by its acronym and AWB Limited tended to abbreviate its name to AWB, blurring the distinction between the old statutory arrangements and the new privatised body. During the debate following the Oil-for-Food scandal, senior ministers continued to refer to the company as the 'Wheat Board', even though by the time of the Cole Inquiry such a body had not existed for 7 years.

It should also be noted that AWB Limited is not the first grain trader to have caused headaches for governments seeking to implement economic sanctions. Conrad writes that

> the primary task faced by the American government in its embargo of grain sales to the Soviet Union in 1980 was to find some way of more closely monitoring and controlling the behaviour of these [private grain trading] companies, most of which handle their Soviet trading through their associated companies in Europe. (Conrad 1990: 118)

Likewise, 'Rhodesia's ability to circumvent international trading sanctions from 1968 to 1975 could not have taken place without the knowledge and assistance of the larger multinational traders' (Conrad 1990: 118). The difficulty for the Australian Government in practical terms was the problem identified above of monitoring compliance with the sanctions against Iraq. In political terms, it was distancing itself from what was essentially a private company but with which it had both close historical links and a contemporary relationship by virtue of the export monopoly.

Oil-for-Food, AWB Limited and Institutional Reproduction

One of the lessons from historical institutionalism relates to the unintended consequences of strategies of institutional reproduction. The Australian Wheat Board had apparently adapted successfully to the removal of its monopoly on the domestic wheat market. It engaged a new type of employee who was engaged in trading in the domestic wheat market, employing sophisticated marketing and risk management strategies and motivated largely by personal ambition to succeed. This new breed contrasted with the former employees of the statutory marketing organisation who were driven by the collective values of the old Wheat Board with its focus on maximising returns to growers while spreading the risk of price fluctuations across all pool participants. This layering of new values over the old was apparently successful in the early days of the deregulated domestic market. However, the new 'traders' became increasingly important in the export trade following privatisation and were clearly susceptible to the incentives to participate in the bypassing of the UN sanctions. As individuals, the traders at the heart of the sanctions evasion were not only keen to hide their activities from the Australian Government but were also, to varying degrees, successful at covering their tracks within their own organisation. There remain senior managers from AWB Limited from that period who are adamant that they were unaware of the kickbacks and the mechanisms being used to pay them.

In 1989 it appeared that the Australian Wheat Board and the institution of collective marketing that it embodied had survived relatively intact. The organisation adapted to the changed economic environment and its new legislative framework and managed to hang on to key features of the pre-1989 arrangements which were most valued by its supporters – grower involvement in the running of the Board, collective sharing of risk through the pools, and the export monopoly which left out any 'middlemen' and prevented Australian wheat from competing on international markets with other Australian wheat. However, the Board's strategies of institutional reproduction and survival contained a major flaw. With the introduction of traders into the organisation and a growing commercial focus following privatisation, there was a widening gap with the values on which the organisation's legitimacy was based and those of the new organisational culture.

Path dependent institutional reproduction can be explained in several ways (Mahoney 2000). First, survival can be explained in purely utilitarian terms in that 'any potential benefits of transformation are outweighed by the costs' (Mahoney 2000: 517). Skogstad (2005) describes a utilitarian process of reproduction in the case of the Canadian Wheat Board; the institution defends its existence in terms of its economic value to prairie wheat growers. Utilitarian explanations of institutional survival emphasise the need for opponents of the arrangements to be able to make the case that the costs of the *status quo* outweigh the benefits. A second explanation emphasises the importance of the institution to the overall functional needs of the system. Mahoney (2000: 519) outlines the functional explanation as follows:

> functionalist logic identifies predictable self-reinforcing processes: the institution serves some function for the system, which causes the expansion of the institution, which enhances the institution's ability to perform the useful function, which leads to further institutional expansion and eventually institutional consolidation. Thus, system functionality replaces the idea of efficiency in utilitarian accounts as the mechanism of institutional reproduction.

A third explanation for institutional path dependence is a power explanation. This involves an elite group of actors who support the institution. Change occurs when there is a shift in power relations within society and the erstwhile opponents of the institution move into a position of influence. This is another way of explaining a shift in the structure of a policy community which allows for alternative views and new players into the policy process.

The fourth of Mahoney's categories is legitimation and this is the explanation that matches most closely with the institutional evolution of the Australian Wheat Board into AWB Limited. Mahoney (2000: 523) argues that: 'In a legitimation framework, institutional reproduction is grounded in actors' subjective orientations and beliefs about what is appropriate or morally correct'. Change occurs as a result of the emergence of a disjuncture between the legitimating values of the organisation and the values that it has come to represent:

> The legitimacy underlying any given institution can be cast off and replaced when events bring about its forceful juxtaposition with an alternative, mutually incompatible conceptualization. Depending on the specific institution in question, the events that trigger such changes in subjective perception and thus declines in legitimacy may be linked to structural isomorphism with rationalized myths, declines in institutional efficacy or stability, or the introduction of new ideas by political leaders. (Mahoney 2000: 525)

Although growers continued to benefit in utilitarian terms from the pooling and export monopoly activities of AWB Limited, the Oil-for-Food scandal revealed the extent to which it was *belief in the organisation and the values embedded in it* that was sustaining support for the institution. Employees of AWB Limited may not have been able to articulate the idea that the legitimacy of their operations and support from industry were tied to the 'old' set of collective values, but they certainly behaved as if they understood this. During the Cole Inquiry, AWB Limited employees continued to use the values language of the early Wheat Board. One senior manager argued that '…I can attest to the fact that my clear intention at all times was to maximise grower returns, look after the interests of the farmer' (Cole Inquiry Transcript: 1686). The same witness subsequently explained the motives for his actions as follows:

> the frame of mind I was probably in was how is this going to impact on grower returns, and that would have been of prime concern to me, and are these contracts going to be executed, are the original pricing forecasts that we put into the pool pricing model—were they still accurate, how did that impact on the estimated pool return we were telling the farmers, and how would that impact on their incomes? Mr Commissioner, always an AWB employee, especially in international sales and marketing— it is always at the forefront of your mind, "How is this going to impact on the grower?" (Cole Inquiry Transcript: 1849)

Within the grains industry, reactions to the scandal were mixed. Public meetings of growers following the revelations of the Volcker and Cole Inquiries indicated that there remained a group of wheat producers who were loyal to AWB Limited and who accepted the kickbacks as the price of doing business in Iraq at that time. For them the imperative of selling the wheat crop into an important export market excused the bypassing of sanctions. The Grains Council of Australia was more wary and by 2005, was refusing to accept payments from AWB Limited which, as discussed in Chap. 7, had become an important part of the Council's operating budget. The industry was particularly worried about the potential fall-out of the revelations for the export monopoly. The media reporting of the scandal had perhaps overplayed the degree to which the Australian Government remained involved in wheat marketing but it drew clear attention to the monopoly arrangements (see for example Sydney Morning Herald 2006).

Conclusion

This chapter described the Oil-for-Food scandal which captured the attention of the Australian nation for much of 2006 as daily newspapers carried detailed reports of the monopoly wheat exporters' bypassing of UN sanctions against Iraq. As a member of the 'coalition of the willing', Australia had troops in Iraq fighting the regime so, to the general population, AWB Limited's activities were almost treasonous. The Federal Opposition capitalised on the daily revelations and the country was entertained by reports of AWB Limited employees going to extraordinary lengths to avoid stating the obvious about their involvement in Iraq. The public gallery for the inquiry grew from a very small number of dedicated journalists on the first day to standing room only. As the public gallery grew so did the number of lawyers and they became increasingly cramped in the seating space allocated for them. When a number of exchanges with witnesses generated laughter from the public gallery, the Counsel for AWB Limited requested the Commissioner to 'ask those who are laughing to desist from laughing. This is not a matter for laughing' (Cole Inquiry Transcript: 2497). When the Tigris matter was the subject of the Inquiry and Mr. Charles Stott was questioned at length about whether the wheat had been donated, Commissioner intervened, stating 'It's a loan, for heaven's sake' (Transcript: 2212). The Counsel Assisting asked another witness, 'How is business done at the AWB if nobody reads their emails, …?' (Transcript: 2859).

For the Opposition in the Federal Parliament, it was a heaven-sent political opportunity. Day after day in Parliament, the leader of the Opposition, Mr Beazley and the Opposition spokesman on Foreign Affairs, Kevin Rudd, went on the attack, asking questions about the Government's knowledge of the activities of AWB Limited, drawing on the evidence coming to light in the public inquiry and seeking to link Ministers to the scandal. Censure motions were moved, and lost along party lines, alleging incompetence, and worse, by the Government in its handling of the scandal. An example of the tenor of the attack is the following by Mr Beazley:

> This is a scandal that goes to Australia's international reputation, the safety of Australian troops serving in Iraq and our country's national security. It is a sorry story of a government in a mode of reckless negligence, now fully exposed by evidence to the Cole inquiry and the shameful revelation that, presented with mounting evidence of kickbacks, the Prime Minister, the Minister for Trade and the Minister for Foreign Affairs all turned a blind eye. (Beazley 2006: 21)

Two Ministers were actually called before the Inquiry but no major revelations emerged that threatened their positions. Mr. Rudd, however, built a reputation and profile which positioned him effectively to take over the leadership of the Labor Party and lead it to victory in the 2007 Federal election. Following that election, the incoming Minister for Agriculture, Fisheries and Forestry made it very clear that the monopoly status of AWB Limited was over (Burke 2008).

References

Australia. Senate Rural and Regional Affairs and Transport References Committee (2005) Iraqi wheat debt— repayments for wheat growers. AGPS, Canberra

Australian Wheat Board (1996) AWB off the mark in Iraq. AWB Grain Statement, 22 May 1996

AWB Limited (undated) Submission to Senate Rural and Regional Affairs and Transport Reference Committee Inquiry into 'compensation arrangements after writing off of the Iraqi wheat debt'

Barber J (1979) Economic sanctions as a policy instrument. Int Aff 55(3):367–384

Bartos S (2006) Against the grain: the AWB scandal and why it happened. UNSW Press, Sydney

Beazley K, the Hon MP (2006) Prime Minister, Deputy Prime Minister, Minister for Foreign Affairs: censure motion. House of Representatives Hansard, 7 February

Botterill LC, McNaughton A (2008) Laying the foundations for the wheat scandal: UN sanctions, private actors and the cole inquiry. Aust J Polit Sci 43(4):583–598

Burke T, the Hon MP (2008) Government takes next steps to end wheat monopoly. Media Release by the Minister for Agriculture, Fisheries and Forestry DAFF08/006B, 6 February 2008

Cole TR, the Honourable AO RFD QC (2006a) Report of the inquiry into certain Australian companies in relation to the UN Oil-for-Food Programme. Commonwealth of Australia, Canberra

Cole TR, the Honourable AO RFD QC (2006b) Report of the inquiry into certain Australian companies in relation to the UN Oil-for-Food Programme. Volume 1: Summary, recommendations and Background. Commonwealth of Australia, Canberra

Cole TR, the Honourable AO RFD QC (2006c) Report of the inquiry into certain Australian companies in relation to the UN Oil-for-Food Programme. Volume 2: negotiations and sales July 1999–December 2000. Commonwealth of Australia, Canberra

Cole TR, the Honourable AO RFD QC (2006d) Report of the inquiry into certain Australian companies in relation to the UN Oil-for-Food Programme. Volume 3: sales, allegations and inquiries January 2001–December 2005. Commonwealth of Australia, Canberra

Conrad JA (1990) The entrepreneurial state: market structure and state trading in wheat. UMI Dissertation Information Service, Ann Arbor, Michigan

Doxey MP (1971) Economic sanctions and international enforcement. Oxford University Press, London

Doxey MP (1996) International sanctions in contemporary perspective, 2nd edn. Macmillan Press Ltd, Basingstoke

Evans G, the Hon QC MP (1990) Australian sanctions against Iraq. Minister for Foreign Affairs and Trade News Release, 6 August 1990

Galtung J (1967) On the effects of international economic sanctions: with examples from the case of Rhodesia. World Polit 19(3):378–416

Grains Council of Australia (1990) 'Hawke stands behind grain industry': GCA News 28 August 1990. Noel Butlin Archives Centre. N158, Box 25 AWB9 3 August 1990–28 August 1990, Canberra

Mahoney J (2000) Path dependence in historical sociology. Theor Soc 29:507–548

Malone DM (2006) The international struggle over Iraq: politics in the UN Security Council 1980–2005. Oxford University Press, New York

Meyer JA, Califano MG (2006) Good intentions corrupted: the oil-for-food scandal and the threat to the U.N. Public Affairs, New York

Overington C (2007) Kickback: inside the Australian Wheat Board Scandal. Allen & Unwin, Crows Nest

Pape RA (1997) Why economic sanctions do not work. Int Secur 22(2):90–136

Renwick R (1981) Economic sanctions. Harvard University Centre for International Affairs, Cambridge

Shovelan J (1990) AM Program. ABC Radio, 8 August 1990

Skogstad G (2005) The dynamics of institutional transformation: the case of the Canadian Wheat Board. Can J Political Sci 38(3):529–548

Sydney Morning Herald (2006) Wheat monopoly beyond reform. 1 December

Volcker PA, Goldstone RJ, Pieth M (2005) Manipulation of the Oil-for-Food Programme by the Iraqi Regime. Independent Inquiry Committee into the United Nations Oil-for-Food Programme, 27 October 2005. http://www.iic-offp.org/documents/IIC%20Final%20Report%2027Oct2005.pdf

Chapter 7
The Aftermath of Oil-for-Food and the Death of an Institution

Keywords AWB Limited • Institutional demise • End of collective marketing • The role of values • Rural policy making

This chapter explores the death throes of collective marketing and reflects on what this means for the values embodied in the institutional arrangements that underpinned wheat policy for nearly six decades. The historical institutionalism literature has many strengths. It is empirically rich, it takes account of that important feature of politics, the allocation within society of values (Easton 1953: 129), and it resists ahistorical analysis. However, scholarship in this field has tended to focus on institutional survival, with some discussion of the origins of institutions. Where it has focused less is on the death of an institution. Pierson (2000, 2004) draws our attention to the lessons to be learnt from focusing on institutional evolution as it unfolds over time and to avoid considering institutional survival as a series of snapshots. The risk with the snapshot approach is that important developments are missed. What may appear to be a successful strategy of survival at one point in time may in fact hide the origins of the future downfall of the institution. This has been a key argument of this book. This case study provides an example of an apparently successful adaptation to change which resulted in the undermining of the very factors on which organisational legitimacy was built. The values driving the business activities within the Wheat Board and its privatised successor resembled less and less those of its greatest supporters. Individual greed and questionable personal ethics clearly played a role in the Oil-for-Food scandal but it seems clear that the move from 'marketer' to 'trader' also contributed to creating an environment in which these activities were able to occur.

This chapter outlines the fallout from the Cole Inquiry and the death of the institution of collective wheat marketing in Australia. In the face of public outcry, reinvigorated critics and the loss of political support, the coalition of supporters of the wheat marketing arrangements was no longer able to protect their deep core values. Many of them continued to defend AWB Limited's actions to the bitter end but, in spite of the efforts of the National Party, they lost the debate.

L.C. Botterill, *Wheat Marketing in Transition: The Transformation of the Australian Wheat Board*, Environment & Policy 53, DOI 10.1007/978-94-007-2804-2_7,

Commissioner Cole summarised the fall-out for AWB Limited of the Oil-for-Food scandal as follows:

> The consequences of AWB's actions [...] have been immense. AWB has lost its reputation. [...] Shareholders have lost half the value of their investment. Trade with Iraq worth more than A$500 million per annum has been forfeited. Many senior executives have resigned, their positions being untenable. Some entities will not deal with the company. Some wheat farmers do so unwillingly but are, at present, compelled by law to do so. AWB is threatened by law suits both in Australia and overseas. And AWB has cast a shadow over Australia's reputation in international trade. (Cole 2006: xi)

The Cole Inquiry provided the media and the opposition parties in the Federal Parliament with 9 months' worth of fascinating material as the evidence of witness after witness was reported and the story of AWB Limited's behaviour gradually unfolded. AWB Limited's tactics in managing the saga were very odd. Rather than seeking to expedite proceedings by offering full cooperation, and therefore get the scandal off the front page of the newspapers, the company withheld relevant documents from the Inquiry. Some papers were released in a piecemeal fashion and others were not made available until the last week of the Inquiry in September 2006. The Counsel Assisting the Commission was so frustrated by AWB Limited's stalling that he suggested that he might need to obtain a search warrant in order to gain access to material. Claims by the company that a number of documents were subject to legal professional privilege were tested in the Federal Court.

Commissioner Cole (2006: 235) argued that 'AWB presented a façade of cooperation with the Inquiry. In truth, it did not cooperate at all'. This only served to prolong the Inquiry and the accompanying political and media attention. A major beneficiary of the scandal was the Opposition spokesman for Foreign Affairs, later Prime Minister of Australia, Kevin Rudd. Mr Rudd raised his public profile considerably, leading the Opposition's attack on the Government, seeking to implicate senior ministers in the scandal and implying that they knew about AWB Limited's activities and chose to turn a blind eye. The Government was criticised for the limited Terms of Reference for the Cole Inquiry which, it was alleged, prevented evidence from emerging that would prove that the Australian Government was aware of AWB Limited's activities in Iraq. *The Australian* newspaper, usually a supporter of Liberal-National Party coalition governments, editorialised in March 2006 that '[Foreign Affairs Minister] Mr. Downer's performance over AWB is unacceptable' suggesting that 'The wheat-for-weapons scandal has claimed its first scalp—Mr. Downer's credibility is crippled' (The Australian 2006). This did not prove to be the case as all Ministers involved in the Cole Inquiry escaped relatively unscathed.

AWB Limited and its employees were not so lucky. Commissioner Cole identified twelve individuals who he believed may have breached elements of Australian law and he recommended the establishment of 'a joint Task Force comprising the Australian Federal Police, Victoria Police, and the Australian Securities and Investments Commission to consider possible prosecutions in consultation with the Commonwealth Director of Public Prosecutions and the Victorian Director of Public Prosecutions' (Cole 2006: lxxxii). At the time of writing, there have been no successful prosecutions arising from the scandal.

However, the fall-out for the company, and the Australian grains industry, was significant. As discussed in Chap. 5, the top two goals growers identified for the restructuring of the Australian Wheat Board in 1995 were retention of the export monopoly and grower control of any restructured body. Both of these were effectively lost in 2008. The end of collective marketing as an institution was accompanied by the collapse of the grains industry's representative body, the Grains Council of Australia, which had played an important supporting role for the statutory marketing arrangements, and subsequently for the export monopoly held by the privatised AWB Limited.

The End of the Single Desk

Immediately following the Australian Government's receipt of Commissioner Cole's Report it took steps to remove the export monopoly from AWB Limited and announced a review of wheat marketing arrangements (Howard 2006). The monopoly had been effectively granted to AWB Limited through the mechanism of the veto which the company had over applications to the Wheat Export Authority from other companies to export bulk wheat. In December 2006 the veto power was transferred to the Minister for Agriculture, Fisheries and Forestry and on 23 December the Minister granted two export approvals to companies other than AWB Limited.

On 12 January 2007, a Wheat Export Marketing Consultative Committee was established to consult growers on the future of the single desk. The Committee held public meetings which were attended by a total of approximately 3,700 people and nearly 1,200 written submissions were received. The Committee set out its main conclusions as follows:

> The Committee found that more than 70 per cent of growers who expressed their views during the consultation process continue to support orderly marketing, most commonly referred to as single desk marketing, in one form or another. However, there is a clear mood for some change.
>
> Only around 20 per cent of growers continue to support the status quo: that is, operation of the single desk under the AWB Limited corporate model. The remaining single desk supporters favour a single desk under different governance arrangements, with most favouring a not for profit, grower owned or controlled entity with the single objective of maximising grower returns. Others supported a strengthened Wheat Export Authority (WEA) fulfilling this role. If it could be achieved, a fully demerged AWB (International) Limited (AWBI) from AWB Limited, owned by growers, would satisfy the majority of those favouring a grower controlled single desk. However, many growers indicate that they believe the possibility of achieving this in a reasonable timeframe, if at all, is unlikely.
>
> Around 20 per cent of growers offering a view support liberalised arrangements for bulk exports, either through multiple licensing systems or full deregulation.
>
> However, support for greater liberalisation of exports in bags and containers is very strong and widespread, provided quality standards are maintained, monitored and enforced. (Wheat Export Marketing Consultation Committee 2007: 1)

The report was presented to the government in March 2007 which decided against its public release. The full report only became available publically after a change of government and the receipt of a Freedom of Information request.[1]

On 22 May 2007, the Government announced that growers had until March 2008 to 'establish their own company, separate from AWB Limited, to manage the single desk' (McGauran 2007). Growers were told that if they did not have a company in place by 1 March 2008 to take over the management of the export monopoly 'the Government reserves the right to introduce its own wheat marketing arrangements' (McGauran 2007). The Prime Minister was clearer in his ultimatum to industry, telling the Parliament:

> If growers are not able to establish the new entity by 1 March next year, the government will propose other marketing arrangements for wheat exports. Let me make this clear to the House. The options available would include further deregulation of the wheat export market. (Howard 2007: 2)

The differences over wheat marketing between the Liberal Party and their coalition partner, the National Party were once again exposed. In spite of this apparently unambiguous indication that deregulation was a real possibility under a Coalition Government, the leader of the National Party in the Senate, Senator Ron Boswell was moved to claim the decision as a victory for the National Party and that the single desk had been 'saved': 'the minister delivered the future of the wheat industry back into the hands of the growers themselves. That is why The Nationals exist. That is why we have a past, a present and a future' (Boswell 2007: 48). A change of government in November 2007 meant that that the Liberal-National Party Government was not faced with dealing with internal division or acting on the Prime Minister's implied threat. The incoming Labor Government very quickly made it clear that the export monopoly arrangements were over (Burke 2008a).

Legislation was introduced into the Australian Parliament in June 2008 to end the export monopoly arrangements. A Senate Inquiry into the legislation was held, attracting submissions from individual farmers, farm organisations, AWB Limited and potential wheat exporters. Remarkably, some farmers continued to downplay AWB Limited's actions in Iraq. Mr. Jock Munro told the Senate Inquiry:

> We all know what really happened with the Cole inquiry and the Iraq oil for food business. It was all about getting food to the starving Iraqi people, and of course there had to be a kickback paid to Saddam Hussein—everybody knows that; it was just that we got caught up in the mess. There were 2,200 companies paying those kickbacks. If it was such a bad thing, why didn't the UN stop it? (Senate Rural and Regional Affairs and Transport Committee 2008: RRA&T 13)

The 2008 legislation did not introduce a completely free market for bulk wheat but established a system of accreditation of potential exporters, giving consideration to issues

[1] http://www.daff.gov.au/agriculture-food/wheat-sugar-crops/wheat-marketing/wemcc

> such as the financial resources available to the company, its risk management systems and the demonstrated behaviour of the company and its executives including making sure that they are abiding by Australian law and complying with foreign laws and United Nations resolutions. (Burke 2008b: 3860)

The Wheat Export Authority became Wheat Exports Australia, responsible for developing and administering the accreditation scheme for bulk wheat exports. The reason given for the accreditation process was 'to make sure that growers are dealing only with companies or cooperatives of good standing and financial capability. Growers need to know that exporters have the reputation and financial backing to pay for their crop'. AWB Limited was not given any special status under the legislation and would 'need to apply for accreditation based on the same criteria as applied to other exporters' (Burke 2008b: 3861). The first round of accreditations was announced in August 2008 (Australian Government. Wheat Exports Australia 2008b). To the consternation of its supporters AWB Limited was not among those listed; it was not until the third round of accreditations that AWB Limited was included (Australian Government. Wheat Exports Australia 2008a).

The accreditation process was only implemented as an interim measure. The *Wheat Export Marketing Act 2008* required the Government's economic review agency, the Productivity Commission, to review the Act and the operation of the accreditation scheme in 2010 with a requirement to report by 1 July that year. Productivity Commission inquiries are transparent, public processes that involve a call for public submissions, the preparation and release of a draft report, the receipt of further submissions plus the holding of meetings at which individuals and organisations can put their arguments. At the end of the process, the Commission presents the government with a final report which the government must table in Parliament within a statutory period. The government then prepares a response; it is not bound by the Commission's recommendations but the Commission is highly regarded as a rigorous, independent economic assessment agency and as such has some sway over government policy. In its final report on the wheat marketing arrangements, the Productivity Commission concluded that

> The accreditation scheme, Wheat Exports Australia and the Wheat Export Charge should be abolished on 30 September 2011. (Australia. Productivity Commission 2010c: 2)

The Commission expressed the view that 'the transition to competition in the marketing of bulk wheat exports has progressed remarkably smoothly and the industry has performed well under the new arrangements' (Australia. Productivity Commission 2010c: 6). The Report acknowledged that the accreditation scheme had 'provided comfort to growers and international buyers in a period of rapid and substantial policy change' (Australia. Productivity Commission 2010c: 10). However it argued that these arrangements were no longer necessary in light of the smooth transition to a deregulated export market. The Commission also noted advice from the Department of Foreign Affairs and Trade that the Doha Round of trade negotiations was likely to result in rules that would have necessitated the changes to wheat marketing arrangements such as those in the 2008 Act, and further that the reintroduction of single desk arrangements with respect to the US would not be possible under the provisions of the Australia-US Free Trade Agreement (Australia. Productivity Commission 2010c: 38). As at December 2011, the Australian Government had not provided a formal response to the Productivity Commission's recommendations so the 30 September 2011 termination date proposed for the accreditation arrangements has not been met.

The End of Grower Control

After the preservation of the single desk, the next most important objective identified by growers during the privatisation debate of the mid-1990s was the entrenchment of grower control of any restructured entity. Growers were concerned that a private company would be more interested in maximising returns to its shareholders than to growers and that this would impact directly on grower returns. As a result of this concern, the model that was adopted introduced two classes of shares, as outlined in Chap. 5. The Oil-for-Food scandal generated concern about AWB Limited's corporate governance and there were questions during the Cole Inquiry as to whether the Australian Stock Exchange had been misled by the company in its denials of wrongdoing (Newman and Korporaal 2006). A class action was commenced against AWB Limited in 2007 on behalf of Class B shareholders who suffered a loss as a result of the drop in the company's share price from over $A6.30 to under $A2.40 in less than a year. The action claimed that the AWB Limited share price would have been lower had the company not concealed its breach of the sanctions against Iraq (Carswell 2007).

The outcome of the Cole Inquiry increased the focus on AWB Limited's unusual share structure and pressure mounted for normalisation of the share register. In a media release in February 2008, the Australian Shareholders Association supported a resolution which would remove the distinction between A and B class shares. The Association was concerned that

> At present A class shareholders can elect 8 directors, and B class shareholders 2 directors. As a result A class shareholders representing wheat growers may act in the interests of growers rather than all shareholders. A class shareholders have no money invested in AWB. If the company performs poorly they incur no loss and yet they elect the majority of directors. (Australian Shareholders Association 2008)

The Wheat Export Marketing Consultation Committee also reported concern among growers about the share structure, but for different reasons. Under a subheading 'conflict in the current corporate structure', the Committee reported that 'The conflict of interest in the AWB Group structure was a primary focus of many growers at meetings. Similarly, the issue is emphasised in grower submissions' (Wheat Export Marketing Consultation Committee 2007: 25). Elsewhere in its report, the Committee reported that 'Growers are particularly concerned about the conflict of interest between AWBI's responsibility for managing the single desk for growers' benefit and AWB Limited's broader set of commercial activities and responsibility to its B Class shareholders' (Wheat Export Marketing Consultation Committee 2007: 8). The solution for many growers was to 'get back to the original

Australian Wheat Board model' (Wheat Export Marketing Consultation Committee 2007: 8). As this course of action was highly unlikely, the alternative approach to normalising the share structure was to remove the final preferential treatment of grower shareholders.

An attempt to change the share structure was defeated in February 2008, receiving nearly 90% support from shareholders overall but failing to obtain the 75% support from A class (grower) shareholders required by the company's constitution. The strong support was interpreted by advocates of reform as a mandate to continue to press for change (AWB Limited 2008a). In September 2008, AWB Limited announced that it had achieved the required support from A class shareholders, effectively ending the era of 'grower control' and putting control of the company into the hands of its owners rather than its customers (AWB Limited 2008b). In July and August 2010, AWB Limited was subject to a great deal of market interest, including a merger proposal from Graincorp, and takeover bids from US based Gavilon and the Canadian based company Agrium. In October, Australia's Foreign Investment Review Board approved the Agrium takeover and AWB Limited recommended that its shareholders accept the offer (AAP 2010). This occurred on 16 November 2010 with 98% of those casting a vote approving the deal at a price of $A1.50. Media reporting speculated that smaller shareholders, including growers, opposed the deal (Urban 2010). On 15 December 2010, Agrium announced that it had sold the AWB Commodity Management business to Cargill, Incorporated. In the media release announcing the sale, Agrium described Cargill as 'one of the world's leading grain handlers and traders' (Agrium 2010). This announcement would have been particularly galling for the growers who attended the meetings in 1995 and who were concerned that a privatised Australian Wheat Board might find its way into the hands of one of the world's big grain companies. The sale received approval from the Australian Government in May 2011. Cargill has indicated that it will be keeping the AWB brand and 'capitalise on its international reputation to secure premium pricing' (Sutton 2011).

In less than a decade after privatisation, AWB Limited lost both the characteristics that had been central to the Grains Council's agreement to the privatisation process— the export monopoly and grower control. The proximate cause of this change was AWB Limited's behaviour in complying with Iraqi schemes to undermine the UN sanctions regime. However, when viewed in a broader historical context, the cultural shifts that occurred within the Australian Wheat Board in response to domestic deregulation in 1989 were perhaps as important in ending 60 years of the wheat export monopoly arrangements.

The Aftermath

The Values Dimension

The debate over wheat marketing in the 1930s and 1940s was resolved in favour of the advocates of collective wheat marketing and for 60 years, the Australian Wheat Board and then AWB Limited embodied these values. Over time, shifting societal values and associated changes in the policy environment put pressure on the institutional arrangements which adapted and responded while protecting the core values of grower control and the export monopoly. The strategies of reproduction adopted in response to these challenges were initially successful but ultimately resulted in the demise of the institution. With the Productivity Commission's Report of 2010, the final vestiges of government intervention in wheat marketing are drawing to a close and a new set of values dominates. The market liberalism which began to influence agricultural policy in the 1970s had already resulted in the deregulation of other key agricultural industries in Australia and it is perhaps remarkable that the wheat industry resisted complete deregulation so successfully for so long. The end of collective marketing has not silenced its supporters. The National Party remains an advocate for the export monopoly. The Party's policy platform is relatively circumspect on the point, stating that 'Where producers want collective export marketing, The Nationals will continue to provide support to establish a fair system in order to maximise grower returns' (The Nationals 2010: 48). In the Parliament, however, its members are more outspoken. In June 2010, before the final Productivity Commission Report had been received by the Government, the Nationals Leader, the Hon Warren Truss MP told the House of Representatives that

> this government sacrificed the wheat single desk on the grounds that it was somehow or other going to lead to better international trade agreements. Australian wheat growers are already paying a very heavy price for the abolition of the Australian organised wheat marketing arrangements. It is already costing this country hundreds of millions of dollars and we have had absolutely nothing in return. There have been no concessions whatsoever from any other country in response to the so-called free trade advantages that there would be as a result of getting rid of the organised marketing arrangements for wheat. What we do know is that Australian growers have lost heavily; the reputation of Australian wheat is in decline; many of our major markets have been lost to countries like Canada, which still have organised marketing arrangements; and in reality the future for the Australian grain-growing industry is quite perilous at present. (Truss 2010: 5628)

On an earlier occasion, in November 2009, his colleague Senator Fiona Nash mounted a passionate case against the removal of the single desk arguing that

> If ever there was a stupid decision by a government, that was it because it has done absolutely nothing but make circumstances worse for the majority of our farmers. Quite frankly, in my view we should have a single desk back. I know we cannot go back to exactly the same old single desk system, but there is no doubt in my mind that we should be going back to an orderly system of marketing wheat, and I believe that because I know that the majority of growers also believe that. (Nash 2010: 8068)

It is not clear that the majority of growers do in fact believe in the old system but there remain vocal advocates. Mr Jock Munro is a wheat grower who is regularly in the media calling for the return to the 1948 arrangements. When the Productivity Commission held public inquiries during the preparation of its 2010 report, Mr Munro told the Commissioners that 'I feel as I'm definitely representing my district and the general rank and file' and went on to mount a spirited defence of collective marketing and stabilisation (Australia. Productivity Commission 2010a: 494–505).

The international trade environment has also changed considerably. There was general consensus in 1948 around intervention in the wheat trade and statutory marketing arrangements were accepted as one mechanism for the successful implementation of the International Wheat Agreements. The advent of the World Trade Organization signalled a major change in the way in which agricultural trade is regarded and the ongoing Doha Round has flagged statutory marketing arrangements as an area in need of reform.

State trading has only recently gained the attention of trade negotiators. When the Australian Wheat Board was created in the late 1940s, state trading enterprises (STEs) were an accepted, and legitimate, feature of the international trading environment (Dixit and Josling 1997: 2; McCorriston and MacLaren 2002: 109), particularly in the area of primary commodities. As Webb (436) notes in his 1953 discussion of the future of international trade

> In the period between the two world wars, prices of primary commodities fluctuated so widely and were in general so low in relation to production costs that State marketing organizations designed to stabilize returns to producers became the rule in almost all countries with large primary industries.

This situation continued into the twenty-first century. McCorriston and MacLaren observed in 2002 (p. 114) that 'with the exception of the European Union, all of the main exporters and importers of agricultural goods use state trading enterprises to manage some elements of their agricultural trade'. As a result of the ubiquity of these arrangements, the General Agreement on Tariffs and Trade (GATT), and the aborted International Trade Organization negotiations, accepted state trading enterprises 'as permanent' but subject to rules 'obliging them to conduct their operations in accordance with the principles assumed to govern the operations of private traders' (Webb 1953: 429). The disciplines on STE activity were included in Article XVII of the GATT and included a requirement on member states to report on the activities of STEs, however not all members with STEs submitted notifications. The provisions of Article XVII remained unchanged by the Uruguay Round of multilateral trade negotiations, although an Understanding on Interpretation of Article XVII was accepted which provided a definition of state trading enterprises and sought to improve compliance with the reporting requirement (Ingco and Ng 1998: 3). In the context of the present discussion about AWB Limited, it is worth noting that the nature of the ownership of a trading entity is not important in determining whether it is reportable as an STE by a GATT signatory but rather it is whether it has been granted special or exclusive rights by government and the impact the exercise of those rights might have in international trade (McCorriston and MacLaren 2002: 110). Australia has therefore been reporting on the activities of AWB Limited as an STE. In 2010, it notified the Working Party on State Trading Enterprises that '**Since AWB (International) Limited no longer holds the single desk rights for bulk wheat exports, it will not be included in future STE notifications**' (WTO 2010: 27 – emphasis in original).

Although the GATT treatment of state trading enterprises did not change significantly between 1947 and the beginning of the Doha Round in 2001, economists and policy makers have become less accepting of their existence. Analysis of STE activity (see for example Ingco and Ng 1998) suggests that these arrangements have the potential to be used to bypass commitments made in other areas of the Uruguay

Round, although the diversity of state trading enterprises 'makes it difficult to generalize about the distortionary effect of STEs and their effects on particular markets or on the world trading system (Ingco and Ng 1998: 9; see also McCorriston and MacLaren 2002). Two of the key features of STEs which have been highlighted as of concern, and were of importance in the former Australian system, are export monopolies and the practice of price pooling. The latter arguably provides STEs with the capacity for greater flexibility than their private competitors due to the delayed payment system associated with pooling (Ingco and Ng 1998: 12).

An important and long standing opponent of STEs, particularly in the wheat trade, is the United States and its particular target has been the Canadian Wheat Board, although US Wheat Associates has also been very vocal about the Australian Wheat Board and AWB Limited. Skogstad (2008: 129) writes of a four-pronged approach by the US against the CWB initiated by then US Trade Representative Robert Zoellick. She describes the four elements of the attack as follows:

> first, working with the US wheat industry to examine the possibility of filing countervailing and anti-dumping cases against Canadian wheat and durum; second, examining a possible WTO case against the Wheat Board, third, working with the US wheat industry to identify specific impediments to U.S. wheat entering Canada; and fourth, 'vigorously pursuing comprehensive and meaningful reform of monopoly state trading enterprises in the [Doha Round] WTO negotiations'.

Between 1989 and 2005 there were 14 challenges by the US against the CWB (Skogstad 2008: 129). In 2002, the US sought consultations with Canada through the Dispute Settlement Body of the World Trade Organization over the collective wheat marketing arrangements embodied in the Canadian Wheat Board (Schnepf 2006). This was followed in March 2003 by a request that a Dispute Settlement Panel be set up. The Panel found that the CWB was not in breach of WTO rules applicable to STEs and this finding was upheld in the face of subsequent US appeals.

Skogstad (2008: 133) argues that the result of the WTO Panel 'confirmed that a monopoly marketing grain agency like the [Canadian] Wheat Board is consistent with the existing international trading regime'. The first two prongs of the US attacks having failed, the Doha Round became the next venue for their approach. As Schnepf (2006: 16) points out

> U.S. wheat producer groups and the USTR remain very disappointed in the WTO's panel ruling with respect to the CWB and are likely to aggressively pursue the elaboration of greater disciplines on state trading enterprises like the CWB in ongoing and future trade negotiations.

In a brief statement to the Second Special Session of the Committee on Agriculture in June 2000, the US listed 'disciplines on single-desk state trading enterprises, import and export' among 8 'key points' in its proposal for agricultural trade reform (WTO 2000). This is one of a number of proposals relating to STEs put forward in the Doha Round. In 2002 McCorrison and MacLaren (p. 129) summarised the most prominent proposals as being those designed

> to measure the trade distorting impact of state trading enterprises; related to this, to categorise state trading enterprises into those that are likely to be 'harmful' and those that are

unlikely to be so; and to place the issue of state trading enterprises as part of the general trade and competition agenda.

By 2008, the multilateral trade negotiations had progressed to the point where STEs were addressed in a separate annex of a 'modalities' document (WTO 2008) prepared by the Chair of the Agriculture Negotiations. The draft proposed the following disciplines on STEs:

> In order to ensure the elimination of trade-distorting practices with respect to agricultural exporting state trading enterprises as described above, Members shall:
>
> (a) eliminate, in parallel and in proportion to the elimination of all forms of export subsidies including those related to food aid and export credits:
>
> (i) export subsidies, defined by Article 1(e) of the Uruguay Round Agreement on Agriculture, which are currently provided, consistently with existing obligations under Article 3.3 of the Uruguay Round Agreement on Agriculture, to or by an agricultural exporting state trading enterprise;
> (ii) government financing of agricultural exporting state trading enterprises, preferential access to capital or other special privileges with respect to government financing or re-financing facilities, borrowing, lending or government guarantees for commercial borrowing or lending, at below market rates; and
> (iii) government underwriting of losses, either directly or indirectly, losses or reimbursement of the costs or write-downs or write-offs of debts owed to, or by agricultural exporting state trading enterprises on their export sales.
> (iv) by 2013, the use of agricultural export monopoly powers for such enterprises. (WTO 2008: 69)

Should these provisions be agreed and adopted in this form, they would have clear implications for the Canadian Wheat Board. With respect to the Australian arrangements, the end of collective wheat marketing has therefore coincided with an important shift in international attitudes towards state trading.

Impact on the Policy Community

An important element of the longevity of the collective wheat marketing arrangements was the strongly supportive role played by the Grains Council of Australia and its predecessor the Australian Wheatgrowers Federation. The relationship was mutually beneficial. The Wheat Board and AWB Limited capitalised on the institutionalised consultation processes that had been built up over time to limit industry criticism and provide a cheer squad for the single desk export arrangements. The career path of wheatgrowers in agri-politics frequently culminated with a seat on the Australian Wheat Board and the entrenchment of grower directors in the privatised body ensured this career option remained available after privatisation. Although some wheatgrower representatives have pursued Parliamentary careers, for example former Cabinet Ministers Warren Truss and Ian Macfarlane, both previously office holders in the Queensland Graingrowers Association, the more common path has been to remain in agri-politics. Whitwell and Sydenham note of members of the Australian Wheatgrowers Federation

> The usual career path was for AWF delegates to stay within the industry rather than make the transition to politics. Prominence in one of the AWF's affiliate organisations might be rewarded by an appointment as a member of the AWB or as a director of a bulk handling authority. (Whitwell and Sydenham 1991: 172)

Before the mid 1980s progress to the Australian Wheat Board could occur without members relinquishing their positions on the grower organisation, prompting the Industries Assistance Commission to observe in 1983, 'The presence of members of the AWF Executive on the AWB may compromise the ability of the AWF to review critically the operation of the AWB. However, this matter is for the AWF to consider' (IAC 1983: 61).

In addition to the promise of a seat on the board, the Australian Wheat Board further consolidated its potential influence over the executive of the grains industry body through the financial arrangements enshrined in the wheat marketing legislation. Successive versions of the *Wheat Marketing Act* included consultation provisions which required the Board to consult regularly with the industry peak body. From 1984, the AWF was named in the legislation and the Board was required to 'consult with the Australian Wheatgrowers' Federation with respect to the performance by the Board of its functions' (*Wheat Marketing Act 1984*, s13). A 1987 amendment to the legislation allowed for the Board to 'reimburse the Grains Council such expenses as the Council reasonably incurs, or has, on or after 1 November 1986, reasonably incurred, in connection with consultation' (*Wheat Marketing Amendment Act (No 2) 1987*, s5). The 1989 Act also allowed for the Board to establish consultative groups beyond the industry body and determine the conditions of remuneration and allowances and meet the expenses of the people selected for consultation. As discussed below, this 'AWB family' became an important source of support for the Wheat Board and, post privatisation, for the single desk.

The 1989 Act also specified that the Board should consult the Grains Council of Australia and, as was the case with the general consultations, the Board was empowered to 'meet expenses reasonably incurred in relation to the consultation by the Grains Council or a representative of the Grains Council' (*Wheat Marketing Act 1989*, s10(3)). As noted earlier, the Grains Council ran its Executive meetings back to back with the Wheat Board-funded consultation meetings, saving itself the costs of travel associated with bringing its executive members together. In addition to the individual disincentive to be critical of the Board provided by promise of future Board membership, the peak body's finances were therefore increasingly dependent on Wheat Board and later AWB Limited funds. These arrangements potentially limited the Grains Council's capacity to be truly independent in its scrutiny of Board activities. In 1995 the Wheat Board carried the bulk of the costs of the 22 grower meetings held during the privatisation debate. This included the chartering of two light planes and their pilots, providing accommodation for members of the GCA Executive and Secretariat as well as Wheat Board staff and financing the dissemination of information prior to the meetings being held. According to an internal GCA memorandum, prior to the round of meetings, the Chairman of the Australian Wheat Board Trevor Flugge threatened to withdraw funding from the GCA if it did not toe the Board's line in the privatisation debate (Grains Council of Australia 1995).

The Australian Wheat Board provided a substantial injection of funds into the Australian Wheatgrowers Federation and its successor the Grains Council of Australia that was in addition to the legislated payments. Former Chief Operating Officer of the GCA, David Ginns, reports that a desk audit of the Grains Council's accounts undertaken at his request in late 2005 uncovered significant payments from the Australian Wheat Board to AWF/GCA as far back as the early 1970s. In the 1980s and 1990s the annual payments from the Board to GCA were around $A250,000 and provided the core of the organisation's budget. Without that money, Ginns suggests, the organisation would not have been viable. The payments continued after privatisation and by the early 2000s the annual payments represented one third of the GCA's Budget. In late 2005, following revelations about AWB Limited's involvement in the Oil-for-Food scandal, GCA, on the initiative of the then Chief Operating Officer and with the agreement of the Council's chairman, ceased invoicing AWB Limited for payment and in February 2006 told the company that the industry body would not accept any further payments (Ginns, personal communication, 2008). There is no evidence of overtly corrupt behaviour by members of the Grains Council relating to these payments but it does raise some questions about the peak industry body's capacity to represent fully the interests of its membership if those interests involved questioning the Wheat Board/AWB Limited's operations.

The Australian Wheat Board not only provided financial support to the GCA. It also provided briefing and policy advice. When the Grains Council organised a delegation to meet with the Prime Minister in August 1990 following the implementation of the initial round of sanctions against Iraq, the Board helpfully provided the GCA's Executive Director with 'Talking Points' for the meeting; presumably to ensure that the industry was convincing in making the case that the Wheat Board needed relief from the financial difficulties resulting from Australia's agreement to the sanctions regime (Grains Council of Australia 1990).

The consultation arrangements with growers outside of the Grains Council provided the Wheat Board and later AWB Limited with an opportunity to coach a cohort of growers in defending the single desk arrangements. During interviews conducted for this research, terms such as the 'AWB Family' and the 'happy clappers' were used to describe growers who participated in the consultation arrangements. Reference was made to an 'AWB hymn sheet' which provided standard answers to criticisms of the Board's operations and arguments to promote the value of the single desk. Interviewees suggested that some of the defenders of the pooling arrangements and the single desk were not in fact delivering their own wheat to the pool as they were obtaining higher prices on the deregulated domestic market. A similar observation was made by the Wheat Marketing Consultation Committee in its 2007 report on the single desk arrangement which noted that inequities potentially arise between those who have access to the domestic market and those who do not (Wheat Export Marketing Consultation Committee 2007: 28).

The collapse of the collective marketing arrangements was quickly followed by the demise of the Grains Council of Australia. Under the leadership of Chief Operating Officer David Ginns, the organisation had already begun the process of moving away from the strongly supporting role that had developed, for example, by cutting its financial ties with AWB Limited following the revelations of the Volcker and Cole Inquiries. After Ginns left the Council in August 2007, it also cut its ties with the National Farmers Federation, apparently for financial reasons. The Chair of the Council, Murray Jones took on the concurrent role as Chief Executive Officer but it appears little activity occurred in the period between Ginns's departure and the subsequent decision to place the Council into voluntary administration in June 2010, although substantial money was spent on the employment of a communications consultant. Former GCA officers involved in the eventual winding up of the organisation report that no new documents were placed on the Council's server after August 2007. Jones retired in July 2009 and was replaced by Jamie Smith as Chairman. An Annual General Meeting of the Grains Council of Australia was held in August 2009 at which the decision was taken to seek renewed support from the grain grower community, or, in the absence of such support, to proceed to close down operations.

Jamie Smith was keen for the GCA to try one last time to seek industry guidance with a view to finding enough common ground within the various grower bodies to see the GCA continue. In October 2009 a Roundtable of 'most of the recognised grower representative groups' (Umbers 2010: 5) was convened to consider the future structure of graingrower representation. Participants agreed that they sought

- A body that is independent and sustainable into the future
- A body that incorporates the existing grower organisations, but can also consider how grass roots engagement can be extended further.
- A body that has the capacity to engage with the wider industry, NFF and other key stakeholders on issues of National importance.
- A body that would provide effective advocacy for Australian growers. (Umbers 2010: 5)

This Roundtable was held to establish the need for and seek input into the nature of a replacement for the Grains Council of Australia, not only to provide a voice for growers in policy debates but also to fulfil the GCA's remaining statutory role with respect to the Grains Research and Development Corporation. The latter is partly funded by levies on growers and there is a requirement in the relevant legislation that the corporation consult with a 'representative organisation' on issues such as the level of the levies payable, and also to report on the research and development activities which the corporation funds. The Act requires the relevant government minister to 'declare one or more specified organisations to be representative organisations', and there must be at least one, for the purposes of the legislation (*Primary Industries and Energy Research and Development Act 1989*, s7). As was the case with the consultative arrangements for the Wheat Board, there are provisions in the Act for some level of funding by the Grains Research and Development Corporation to cover the costs incurred by the representative organisation in participating in the statutory consultations.

Participants at the October 2009 Roundtable agreed to commission independent consultants to prepare a plan for the operations and funding of a body to continue the roles of the GCA. A second roundtable on 17 February 2010 was presented with a business plan which was approved by all present. Participants decided that in the interests of being seen to make a break from the old Grains Council it was important that the industry needed a fresh start without the involvement of people associated with the GCA (Umbers, personal communication, 2010). To that end, an independent Implementation Committee was established to give effect to the business plan, however, by June 2010 this group had released its own strategy document that differed in several important respects from the Business Plan agreed the previous February. A key difference related to the membership structure. The February 2010 Business Plan retained a reduced role for state farming organisations and other groups (Umbers 2010: 18) whereas the June plan indicated that it was seeking membership of the new Grain Producers Australia solely by individual grain growers:

> GPA membership is open directly to grain producers across Australia and is voluntary. Grain farmers can participate in membership of GPA regardless of any other affiliations or non affiliations. GPA recognises collective representation of grain farmers in the industry, but does not provide a membership category beyond grain producers. (Grain Producers Australia Steering Committee 2010: 6)

By this time, the Grains Council of Australia was faced with a deteriorating financial position, and the Board had to consider options including closing the organisation down. Two courses of action were available to the GCA in ceasing operations. It could wind up and liquidate its assets or it could go into voluntary administration and restructure, emerging in a different form. In either case the GCA constitution allowed for the transfer of its assets to a 'like organisation'. As noted above, the Committee charged with implementing the Plan agreed by industry participants in February 2010 had morphed into a new grower body and had discarded the agreed Plan. In July 2010, a proposal was published to establish formally the new body under the name Grain Producers Australia (GPA) along the lines outlined in their alternative June Plan. The first reason given for the need to establish GPA was that 'The Grains Council of Australia's announcement of its financial stress and possible closure presents a huge risk to the industry given its role as the declared representative organisation under several pieces of Commonwealth legislation as well as other non legislated roles'. The vision of the new organisation was set out as follows:

> Grain Producers Australia will be recognised as the pre eminent agricultural advocacy group in Australia with a stable membership representing at least eighty percent of grain growers and ninety percent of grain production. (Grain Producers Australia Steering Committee 2010: 5)

The GCA Board chose to place the company into voluntary administration, and the newly formed Grain Producers Australia assumed the company number, meaning that it was legally the same entity but with a new constitution and board.

The statutory roles that had been granted to the Grains Council were then assumed by the new body.

Grain Producers Australia purports to be a grain farmer advocate representing the production sector of the grains industry. However, the founding Board currently appears to consist only of the Implementation Committee arising from the second Roundtable of February 2010, plus a small number of invited individuals known in the industry or with some business skills. Although the process undertaken by the Grains Council of Australia in winding down its activities was chosen with a view to transferring its statutory roles to the new body as presented to the industry in February 2010, Grain Producers Australia's status remains unclear. The organisation's membership remains very small and it appears to be having difficulty in attracting the direct grower membership it seeks. Industry insiders report that the Australian Government has only conditionally accepted the GPA's claim to be the grains industry's representative organisation and has asked the body to demonstrate that it is truly representative. In the meantime, consultative meetings between industry and the Grains Research and Development Corporation are reportedly including the GPA but also grains industry members of the New South Wales Farmers' Association and other grower groups.

In April 2010, the NSW Farmers Association and the Western Australian Farmers Federation set up a competing structure, National Grains Australia, announcing that they had signed a Memorandum of Understanding 'relating to the future representation of grain growers in Australia' (WAFarmers and NSW Farmers Association 2010). The constitution for this alternative organisation includes among its objectives

- 'establish and thereafter maintain the Company as the recognised peak national representative body for Australia's grain growers';
- 'represent and promote Australia's grain growers, the policies of the Company and the interests of Australia's grain industry nationally and internationally'
- 'maximise the economic and social welfare of Australia's grain growers'; and
- 'fulfil such duties and responsibilities that are from time to time reserved for the Company by, or under, the Government of Australia, or any other Commonwealth or State agency or authority'. (National Grains Australia 2010: 6)

This latter point suggests that this organisation is also seeking to take on the Grains Council of Australia's statutory roles and there is little sign of compromise between the two to arrive at a single organisational structure (Heard 2010). At the time of writing neither organisation can claim to be truly representative of Australian grain growers. The GPA proponents have secured the legal responsibilities of the old GCA with respect to the setting of levies etc, however, their individual membership model is ambitious and it is not clear that it will be sustainable. As part of its desire to have a 'fresh start', the GPA decided it did not want to take ownership of the GCA files so it is operating without the benefit of access to the organisation's corporate history. Fortunately for future researchers interested in this part of the grains industry's story, the process started by Ginns to donate Grains Council papers to the Noel Butlin Archives at the Australian National University was continued as the Council was wound up; in preference to the alternative approach of destroying the records.

The NGA proposal has the support of the Western Australian and New South Wales farm bodies but is unlikely to attract support from the other grain growing states as the voting structure proposed, based on volume of production, would disenfranchise the latter in the event that WA and NSW farmers voted as a bloc. In all respects the industry appears to have returned to the situation that prevailed before the establishment of the Australian Wheat Board when the industry could not agree on a single entity for industry representation.

The demise of the collective wheat marketing arrangements has therefore seen the collapse of an important complementary organisation which had a key role in supporting the old Wheat Board and advocating for its underlying values. It is not entirely clear in the new, deregulated market whether a strong advocate for the grains industry is necessary. One option for the industry is to revisit an option originally on the table in the early days of the National Grain Marketing Strategic Planning Unit to establish an organisation that represents the whole grains industry 'from paddock to plate'. In an important development which would make such a body more likely, the National Farmers Federation in 2008 announced a change to its structure to embrace associate membership by 'agricultural entities' including agribusinesses. Graincorp, the privatised NSW bulk handler is one such associate member. In announcing the change, the NFF's President, David Crombie stated that

> the reformed structure is recognition of the changing face of Australian agriculture. It is becoming far more integrated and the desire and need of farmers to be engaged with others through the supply chain to take forward agriculture's case on national issues is simply a reality… and it's a positive reality. (National Farmers Federation 2008)

A second option is that the Grains Research and Development Corporation, which is funded partly by grower levies, take on the roles of marketing and industry representation role. This has already occurred in some sectors of Australian agricultural industry, notably meat, pork and eggs, and these organisations claim that considerable savings have accrued to their members in terms of reduced corporate overheads (Australia. Productivity Commission 2010b: 182). In a draft report on the rural research and development corporations, the Productivity Commission fell short of recommending the corporations take on these additional functions but did leave the way open for this to occur in the future (Australia. Productivity Commission 2010b: 185). Were this to occur, there would be real questions as to the value of a separate body in either the form of the GPA or the NGA.

Conclusion

Collective marketing in Australia is not only dead as an institution but it has also lost its institutional support structures. While there remain strong advocates of 'orderly marketing' among growers, particularly with respect to an export monopoly and grower control of marketing arrangements, the debate has moved on. The domestic policy environment has long favoured deregulation as the Productivity Commission's 2010 Report makes clear and the international policy community is also considerably less sanguine about state trading arrangements than it has been in the past. With the sale of AWB Commodity Marketing to Cargill, all that is left of a once venerable institution is its brand. All of the other features of the collective marketing arrangements it represented have disappeared from the Australian grains industry landscape.

References

AAP (2010) FIRB clears Agrium takeover of AWB. Sydney Morning Herald, 4 October 2010
Agrium (2010) Agrium sells AWB commodity management business to Cargill. News Release, 15 December 2010
Australia. Productivity Commission (2010a) Draft report on wheat export marketing: transcript of proceedings. Sydney, Tuesday 11 May 2010
Australia. Productivity Commission (2010b) Rural research and development corporations: draft report. Canberra, September 2010
Australia. Productivity Commission (2010c) Wheat export marketing arrangements. Productivity Commission Inquiry Report, No. 51, Canberra, 1 July 2010
Australian Government. Wheat Exports Australia (2008a) WEA accredits AWBL subsidiaries for bulk wheat exports. Media Release, 11 September 2008. http://www.wea.gov.au/Media/Media_Releases/11_September_2008.html
Australian Government. Wheat Exports Australia (2008b) WEA accredits five bulk wheat exporters. Media Release, 26 August 2008. http://www.wea.gov.au/Media/Media_Releases/26August2008MediareleaseAccredit1.html#
Australian Shareholders Association (2008) AWB on the right track. Media Release, 8 February 2008. http://www.asa.asn.au/MediaRelease.asp?ID=MR58.xml
AWB Limited (2008a) ASX: shareholder mandate for AWB reform. Company Announcement, 12 February 2008. http://www.awb.com.au/investors/companyannouncements/current/ShareholdermandateforAWBreform.htm
AWB Limited (2008b) AWB Constitutional reform approved by shareholders. Company Announcement, 3 September 2008. http://www.awb.com.au/investors/companyannouncements/current/AWBConstitutionalreformapprovedbyshareholders.htm
Boswell R, Senator (2007) Wheat Marketing Amendment Bill 2007: Second Reading Debate. Senate Hansard, 21 June 2007
Burke T, the Hon MP (2008a) Government takes next steps to end wheat monopoly. Media Release by the Minister for Agriculture, Fisheries and Forestry DAFF08/006B, 6 February 2008
Burke T, the Hon MP (2008b) Wheat Export Marketing Bill 2008: Second Reading Speech. House of Representatives Hansard, 29 May 2008
Carswell A (2007) AWB share action—investors sue over collapse in value. The Daily Telegraph. 14 April 2007
Cole TR, the Honourable AO RFD QC (2006) Report of the inquiry into certain Australian companies in relation to the UN Oil-for-Food Programme. Volume 1: Summary, recommendations and background, Commonwealth of Australia, Canberra
Dixit PM, Josling T (1997) State trading in agriculture: an analytical framework. Working Paper #97-4, International Agricultural Trade Research Consortium, Washington DC
Easton D (1953) The political system: an inquiry into the state of political science. Alfred A Knopf, New York

Grain Producers Australia Steering Committee (2010) Grain producers Australia. http://www.gpau.com.au/documents.html. Accessed 19 October 2010
Grains Council of Australia (1990) Letter from Ron Storey to Laurie Eakin 20 August 1990. Noel Butlin Archives Centre. N158, Box 25 AWB 9 August 1990–July 1993, Canberra
Grains Council of Australia (1995) Discussion between Greg Ellis and Trevor Flugge: memo from Greg Ellis to GCA Executive 10 October 1995. Noel Butlin Archives Centre. N158, Box 20 GE2.1 Part 1, Canberra
Heard G (2010) Grains bodies set for head-on collision. Stock & Land, 21 May 2010
Howard J, the Hon MP (2006) Joint Press Conference with the Hon Mark Vaile MP, Deputy Prime Minister and Minister for Transport and Regional Services, Parliament House, Canberra. Transcript, 5 December 2006
Howard J, the Hon MP (2007) Questions without notice: wheat. House of Representatives Hansard, 22 May 2007
IAC (Australia. Industries Assistance Commission) (1983) The wheat industry. Report No. 329, Australian Government Publishing Service, Canberra, 29 September 1983
Ingco M, Ng F (1998) Distortionary effects of state trading in agriculture: issues for the next round of multilateral trade negotiations. Policy Research Working Paper 1915, World Bank
McCorriston S, MacLaren D (2002) State Trading. the WTO and GATT Article XVII. World Econ 25(1):107–135
McGauran P, the Hon MP (2007) Interests of wheat growers prevail. Media Release by Federal Minister for Agriculture, Fisheries and Forestry DAFF07/064PM, 22 May 2007
Nash SF (2010) Adjournment: wheat exports. Senate Hansard, 17 November 2009
National Farmers Federation (2008) NFF restructure gets full support of members for '09 launch. Media Release, 18 December 2008. http://www.nff.org.au/read/1595.html
National Grains Australia (2010) Constitution of National Grains Australia Limited. http://www.nationalgrains.org.au/contact_us. Accessed 19 October 2010
Newman G, Korporaal G (2006) Watchdog on alert as AWB changes its tune. The Australian, 20 January 2006
Pierson P (2000) The limits of design: explaining institutional origins and change. Governance 13(4):475–499
Pierson P (2004) Politics in time: history, institutions, and social analysis. Princeton University Press, Princeton
Schnepf R (2006) U.S.-Canada wheat trade dispute. CRS Report for Congress Congressional Research Service, Library of Congress. http://www.opencrs.com/document/RL32426/2006-04-19/download/1005/. Accessed 19 October 2010
Senate Rural and Regional Affairs and Transport Committee (2008) Reference: Wheat Export Marketing Bill 2008 and Wheat Export Marketing (Repeal and Consequential Amendments) Bill 2008 [Exposure drafts]. Official Committee Hansard, 22 April 2008, Official Hansard Report, Canberra. http://www.aph.gov.au/hansard/senate/commttee/S10724.pdf
Skogstad G (2008) Internationalization and Canadian agriculture: policy and governing paradigms. University of Toronto Press, Toronto
Sutton M (2011) Cargill raises stakes for SA grain chain. Stock Journal, 12 May 2011
The Australian (2006) Editorial: Credibility Crippled. 29 March 2006
The Nationals (2010) Policy platform. http://www.nationals.org.au/Policy.aspx, Accessed 19 October 2010
Truss W, the Hon MP (2010) Appropriations Bill (No. 1) 2010–2011: consideration in detail: Foreign Affairs and Trade portfolio. House of Representatives Hansard, 16 June 2010
Umbers A (2010) National grower representation—operational and funding plan. Consideration for the business plan for a national grower body, Umbers Rural Services Pty Ltd. http://www.grainscouncil.com/html_docs/businessplan.html. Accessed 19 October 2010
Urban R (2010) Investors approve Agrium's bid for AWB. The Australian, 17 November
WAFarmers and NSW Farmers Association (2010) National grains Australia proposal. Joint Media Release, 20 April 2010. http://www.nationalgrains.org.au/contact_us
Webb L (1953) The future of international trade. World Polit 5(4):423–441

Wheat Export Marketing Consultation Committee (2007) Wheat Export Marketing Consultation Committee Report. Commonwealth of Australia, Canberra. March 2007

Whitwell G, Sydenham D (1991) A shared harvest: the Australian wheat industry, 1939–1989. Macmillan Australia, South Melbourne

WTO (World Trade Organization) (2000) Second special session of the committee on agriculture: statement by the United States, G/AG/NG/W/32. Committee on agriculture special session. 8 December 2008

WTO (World Trade Organization) (2008) Revised draft modalities for agriculture. TN/AG/W/4/Rev.4, Committee on agriculture special session. 8 December 2008

WTO (World Trade Organization) (2010) State Trading: New and Full Notification Pursuant to Article XVII:4(a) of the GATT 1994 and Paragraph 1 of the Understanding on the Interpretation of Article XVII: Australia. G/STR/N/13/AUS, Working Party on State Trading Enterprises, 13 September 2010

Chapter 8
Lessons and Reflections

Keywords Values • Politics • Institutions • The problem with economic sanctions • Lessons for privatising a monopoly

In 2011 there remain few vestiges of the collective wheat marketing arrangements which were put in place with the establishment of the Australian Wheat Board in 1948. Wheat pools continue to be run by various private companies such as Graincorp, but the ethos of these bodies is based in concerns about shareholder value. There is no longer a single desk for export. The Australian Government has yet to respond to the Productivity Commission recommendation of 2010 that the remaining regulatory arrangements around wheat export be removed. However, the arrangements that do remain have largely 'provided comfort to growers and international buyers in a period of change and financial instability, and facilitated a smooth transition to the deregulated environment' (Australia. Productivity Commission 2010b: 119) rather than provide any level of real governmental control over exports or intervention in the wheat market.

The preceding chapters have sought to speak to two audiences. For the reader interested in the Australian wheat industry, the story has been presented, first, to fill some important gaps in the poorly reported and poorly understood events around the Oil-for-Food scandal by providing some historical context. The sanctions regime against Iraq did not commence with the Oil-for-Food program in 1996, it had been in place since 1990 and Australian graingrowers had already been asked to wear the cost of the original imposition of those sanctions. It was not until 2005 that the issues arising from the Iraqi debt from 1990 were finally resolved. During the Cole Inquiry, the issue of the Iraqi debt was raised and was described by the AWB Limited Chairman as 'an issue that had been bubbling away in the grower politics' (Cole Inquiry 2006: 4027 (6 March)).

The second issue for this reader is related to the first; the international grain industry is highly concentrated in the hands of a few private companies and as a result industry operations are opaque at best. Although it was appealing for the

L.C. Botterill, *Wheat Marketing in Transition: The Transformation of the Australian Wheat Board*, Environment & Policy 53, DOI 10.1007/978-94-007-2804-2_8,

media and the then Opposition to suggest that middle ranking officers in the Department of Foreign Affairs and Trade (DFAT) should have picked up the 'padding' of prices in the contracts between AWB Limited and Iraq, it demonstrated a lack of understanding of the complexity of the international grains trade.

The third point relates to the strength of the rural policy community in Australia. Other privatisations occurred in Australia throughout the 1980s and 1990s but none seemed to slip under the radar to the same extent as the privatisation of the Australian Wheat Board. The usual players in a privatisation, for example the Department of Finance, were not part of the process which was managed by industry, predominantly the Grains Council of Australia and the Wheat Board itself, the very entity being privatised. The rather peculiar company that resulted also managed to slip through the National Competition Policy legislative review process with, again, the review not being conducted by the usual bodies, but by an 'independent' inquiry (Irving et al. 2000), the membership of which included a former President of the Grains Council. I have written elsewhere (Botterill 2006) about the low level of engagement with and critique of rural policy by the Australian public and media. The public becomes interested in rural policy issues when there is some form of scandal, such as Oil-for-Food or, more recently, the revelations of cruelty towards live cattle shipped from Australia to Indonesia (AAP 2011). Substantive policy debates, however, take place within a tight policy community with shared values and from which dissenting voices are largely excluded.

The second audience for this book is the political scientist or public policy scholar. For this reader, the book seeks to achieve a number of objectives. First, it sets out to extend the historical institutionalist literature by considering an institution over its entire life cycle, from cradle to grave. As Rayner (2009: 88) has pointed out 'Exogenous shocks certainly happen and sometimes have the ability to pose urgent new problems, but more often the system contains within itself the seeds of its own destruction'. The case of collective wheat marketing demonstrates that those seeds can be sown as a result of the strategies an institution adopts to respond to change. This is not a particularly new idea that organisational actions can have unintended consequences. For example, Merton (1936: 903) described as an 'essential paradox of social action' the fact that 'the "realization" of values may lead to their renunciation'. However, the focus of the historical institutionalist literature on stability and strategies of reproduction has, in my view, overlooked or downplayed the 'dark side' (Vaughan 1999) of organisations that can produce sub-optimal outcomes. I have argued that it was the 1989 changes to the domestic wheat market, not the more obvious events of the Oil-for-Food program, that ultimately led to the end of the single desk. The coincidence of the timing of the introduction of Oil-for-Food and the privatisation of the Wheat Board no doubt contributed to the events as did the actions of individuals within AWB Limited. The 1989 changes were not the sole cause of the demise of the institutions of collective marketing; as Merton (1936: 898) notes 'the factors involved in unanticipated consequences are – precisely, factors … none of these serves by itself to explain any concrete case'.

A further objective with this second reader in mind was to draw attention to the central role of values in this particular institutional history. The policy communities literature highlights the role of shared values in the policy process and the defence of the single desk by AWB Limited, the Grains Council and individual farmers was conducted in values terms, drawing on agrarian collectivist imagery – even after the Board had been privatised.

This book has sought to tell the story of collective wheat marketing in Australia. It has done this through the lens of historical institutionalism in order to illustrate that the demise of collective wheat marketing was not an accident but that changes to the organisation over time contributed to its downfall. The framework employed is however a 'weaker' institutionalist approach (Rayner 2009: 89) as it acknowledges the important role of individual actors within the organisation in contributing to the events that resulted in the Oil-for-Food scandal. Institutionalists have made an important contribution in the past three decades by refocusing attention on the importance of institutional structures in shaping political behaviour. They have considered the role of institutions as the embodiment of important societal values, and as the result of values compromises reached at a single point in time. They have also examined how these institutions have evolved in response to changing social and economic circumstances and survived through various mechanisms of reproduction.

This chapter seeks to draw these various threads together as follows. The next section explores in more detail the issue of the role of agents in this story and how this understanding fits within the historical institutionalist approach. This is followed by a discussion of the issues of values and their role in sustaining the wheat marketing arrangements and the close link with the grower groups which supported them. Finally, the chapter considers some more prosaic lessons from this story in terms of policy relating to the process of privatising the Australian Wheat Board and the challenges of making economic sanctions work.

The Role of the Individual in the Demise of the Institution

The rise of the 'new institutionalism' in political science in the 1980s is frequently explained as a reaction to the behavioural theory and rational choice perspectives which placed the actor centre stage. The debate around this reaction is often portrayed in terms of a dichotomy between structure and agency which, as Peters and Pierre argue, is a 'useful analytical device, but it should not be taken quite as seriously as it sometimes is in the literature' (Peters and Pierre 1998: 566). Others argue that the perspectives of, for example, rational choice and institutionalism, are in fact not incompatible (Dowding 1994). Putting the debate aside, it is clear that actors within organisations are constrained by organisational values, norms and rules. However, they also have potential to influence, and at times pervert (Granovetter 1985; Vaughan 1999), those values, norms and rules. As March and Olsen point out in one of the first articles to use the term 'new institutionalism', demography matters:

'institutions are driven by their cohort structures, and the pursuit of careers and professional standards dictates the flow of events' (March and Olsen 1984: 744).

In the case of collective wheat marketing in Australia, a cohort of employees was introduced into the Australian Wheat Board in 1989 which was pursuing a career path as wheat traders; as distinct from the wheat marketers of the pre-1989 environment. Their behaviour during the Oil-for-Food program, while arguably unethical, was not illegal and was consistent with the objective of selling the Australian wheat crop for the best possible price. When the details of the trade with Iraq came to light, it triggered the sequence of events that resulted in the removal of the last core element of the 1948 arrangements, the export monopoly. As discussed below, it can only be speculated whether other forces would have seen the removal of the single desk but the manner in which it was removed was the result of the acts of individuals within the institution, not of the structure itself. There is no evidence to suggest one way or another whether the senior management of AWB Limited knew of or condoned the actions of the key players in the kickback arrangements but the lengths to which these players went to cover their tracks suggest that they were acting of their own volition and not under instruction.

Apart from the role of the AWB Limited employees in bypassing the economic sanctions against Iraq, there are other examples in the history of collective wheat marketing which suggest that actors were influential in the direction of institutional development. The privatisation of the statutory Wheat Board was driven by the Board itself which saw advantages in the removal of governmental controls over its activities but clear market value in the retention of the export monopoly. As I have noted, individuals within the Wheat Board could no doubt see the potential personal benefits of privatisation associated with their salaries being set by the market rather than in accordance with public service procedures. The body purportedly representing the graingrowers who would be directly affected by the change was arguably compromised by it close financial relationship with the Board and the personal interests of the GCA Executive for a long time had not always coincided with those of its membership. A career path which on many occasions culminated in a seat on the Board could be cut short by excessive scrutiny of or dissent from Wheat Board directions. This tension between the growers' interest and the grower's interest provides another perspective on the role of individual interest in institutional evolution. A completely disinterested GCA Executive might have fought harder for a corporatised model, with or without the novel 51% grower ownership proposal. The Council's quick capitulation in the face of advice that this latter model was not workable suggests the organisation was inclined to toe the Wheat Board line.

Values, Politics and Institutions

Australia's wheat marketing arrangements had their origins in a form of agrarian collectivism; the belief that farmers needed to pool their crop in order to avoid exploitation by grain traders and that a single exporter was required to ensure that all farmers received the best price for their wheat. The Fair Average Quality (FAQ) classification used for the wheat crop for many years reflected this sentiment and an underlying belief that all growers should be paid the same for their crop irrespective of its quality, distance from port or variety. Like many developed Western nations, Australia has a strong vein of agrarianism running through its culture which has influenced rural policy settings as well as the nature of rural policy debate. With the deregulation of agricultural markets from the 1970s, agrarianism was not quite as explicitly influential over policy settings but it continued to be tapped into by policy makers as they provided rationales for various policy settings; particularly if the policy in question appeared contrary to broader economic policy direction (Botterill 2009). Governments and farm groups appear to get away with these digressions from overall policy direction because there remains within the broader Australian community a strong sympathy for farmers and rural communities. Analysis of a 2009 poll of attitudes towards rural life and agricultural practices and policies suggested that there remain four strands of agrarianism amongst the Australian population: that primary industries are the economic foundation of the nation; that farming as an activity is morally superior; that agriculture faces particular challenges that warrant some level of government intervention; and that there are important locational disadvantages associated with rural life such as reduced access to services. This sentiment inclines the public towards the view that farmers are deserving of government support. This has implications for the level of critique of government rural policy and is a plausible explanation for the low level of interest in issues such as the Australian Wheat Board privatisation.

Against the backdrop of general public indifference to wheat industry policy and politics, the Australian Wheat Board and later AWB Limited were able to tap into the agrarianism of much of their support base to establish the organisation's legitimacy. As described throughout this book, values have an important role in the establishment and preservation of institutions. The original Australian Wheat Board represented the triumph of agrarian collectivist values over those of the free market and the 1948 legislation gave form to that win for farmers who had become suspicious of the 'middle man', particularly during the hard years of the 1930s. For the first four decades, the Board promoted the collectivist values with some incremental changes such as the increased differentiation of wheat varieties but generally the values base of the organisation was secure. I have argued that the turning point was 1989 when the prevailing economic climate of deregulation finally caught up with the Australian wheat industry and the domestic market was opened to competition. From this point, the Board continued to appeal to the values on which it was established, but there is a question about the degree to which they subscribed to them beyond their rhetorical appeal.

The word 'values' is as yet imprecisely defined in the political science/public policy literature. In a classic work on the policy sciences, Lasswell (1951: 9–10) suggests that a value is

> 'a category or preferred events' such as peace rather than war, high levels of productive employment rather than mass unemployment, democracy rather than despotism, and congenial and productive personalities rather than destructive ones.

In much of the literature 'values' is used interchangeably with 'ideas'. This is not very helpful. It would seem to be important to make the distinction between the two as 'There is a normative element to values which distinguishes them from ideas; values are often deeply held and difficult to shift, representing fundamental understandings about the way the world does and should operate' (Daugbjerg and Botterill 2011). An example of the blurring of the two terms occurs in Doern and Phidd's typology of the normative influences on policy (Doern and Phidd 1983: 54). The authors list 'dominant ideas' at the meso level of their schema, locating them between ideologies and policy objectives. As examples of such dominant ideas they include liberty, redistribution and equality, equity, and national identity, unity and integration (Doern and Phidd 1983: 54). They also include efficiency which Etzioni (1988: 245) notes is a value like any other.

These would seem to be values rather than simply ideas which imply susceptibility to change in the light of new information. Ironically, in an article which seeks to provide conceptual clarity around 'ideas' Campbell (1998) uses the term in contexts in which 'values' would probably be more accurate. He proposes a four cell grid which classifies ideas in terms of whether the ideas (which he uses interchangeably with concepts and theories) are in the foreground or background of the policy debate, and whether ideas operate at a cognitive or normative level. He argues that 'at the normative level ideas consist of values and attitudes' (Campbell 1998: 384). This study has taken an approach closer to that of Lasswell, regarding values as guiding principles that are anchored in our particular life experience, and our learnt morality. These values are generally very hard to shift; they are not susceptible to rational argument or the presentation of 'evidence'. They are essentially Sabatier's 'deep core values' which he describes as 'fundamental normative and ontological axioms which define a person's underlying personal philosophy' (Sabatier 1988: 144). In this interpretation, ideas would sit underneath values in a hierarchy of influence, with the cognitive sitting beneath the normative.

If this definition of values is accepted, it raises an important point about the role of values in sustaining an institution. In his typology, discussed in Chap. 6, Mahoney (2000: 523) refers to legitimation as an important strategy of institutional reproduction. He argues that

> In a legitimation framework, institutional reproduction is grounded in actors' subjective orientations and beliefs about what is appropriate or morally correct. Institutional reproduction occurs because actors view an institution as legitimate and thus voluntarily opt for its reproduction.

This explanation does not go to the point of whether the institution is worthy of such loyalty and whether it continues to be appropriate. It would appear that this is where the symbolic appeal to values can be important. The Australian Wheat Board built around it a support network which was invested in the collective values that underpinned its establishment. After 1989, the internal dynamics of the organisation changed and, as evidenced by the Board's positioning during the privatisation debate, its leadership became focused on becoming an important and powerful player in the international grain trade rather than simply the marketer of the growers' crop. However, the appeal to the rhetoric of its founding values continued. March and Olsen draw attention to the potential for a mismatch between values used in public pronouncements and those actually driving behaviour. They argue that

> Symbols permeate politics in a subtle and diffuse way, providing interpretive coherence to political life. Many of the activities and experiences of politics are defined by their relation to myths and symbols that antedate them and that are widely shared. (March and Olsen 1984: 744)

They go on to observe that 'Individuals and groups are frequently hypocritical, reciting sacred myths without believing and while violating their implications'. Brunsson and Olsen (1993: 9) make a similar point that organisations may 'develop double standards, one ideology for internal and one for external use'. This clearly occurred during the Cole Inquiry into AWB Limited's activities during the Oil-for-Food program. Witnesses continued to appeal to the legitimating values of the institution by reference to the growers' interest and to some extent this was successful in retaining the support of many growers who were willing to excuse the kickbacks. Campbell (1998) has argued that 'historical institutionalists ignored how the content of underlying norms and values provides the symbols and other elements that political actors use in carrying out … explicit and deliberate manipulations'. He is referring here to the use by elites of appealing discourses and symbols in order to be persuasive about a particular policy solution. In the case of AWB Limited, the symbols were used in an effort to retain legitimacy in the eyes of the grains industry which provided an important support structure to the institution.

The historical institutionalism literature recognises the role of complementary institutions that arise around a primary institution and provide a supportive environment through shared values and shared understandings of policy problems (see for example Pierson 2004: 27). As I have outlined, the Grains Council of Australia, and its predecessor the Australian Wheatgrowers Federation, were heavily invested in the success of the Australian Wheat Board and its founding values of agrarian collectivism. Their role was institutionalised through various pieces of legislation which allocated formal consultative roles to the Council and this facilitated the development of the close, and highly dependent, financial relationship between the Council and the Wheat Board. This cosy arrangement was sustained for nearly six decades. The Grains Council was a strident opponent of the removal of the monopoly on the domestic wheat trade in 1989 and it was also an important conduit to government of Wheat Board concerns about the implementation of the first sanctions against Iraq in 1990. A delegation from the growers' representative body concerned about the potential losses to individual farmers arising from the sanctions was likely to be more politically effective than a statutory government authority raising the same issues. The Board just made sure that the Council ran the argument effectively by providing 'Talking Points' for those meeting with the Prime Minister.

As discussed in Chap. 7, the collapse of the collective wheat marketing arrangements was followed quickly by the collapse of the Grains Council of Australia. Beyer has noted that, in the event of change, 'all institutions linked by complementarity will come under pressure for adaptation. In the process of transition, "domino effects" may occur' (Beyer 2010: 7). The wheat industry policy community is now closer to the looser 'issue network' end of the Marsh-Rhodes continuum. These networks are larger than closed policy communities; there is a range of interests and values involved; the network is fluid in terms of membership, access and level of interaction between members; and there are unequal resources and power (Marsh and Rhodes 1992: 251). There is no clear successor to the Grains Council as the single voice of the grains industry, leaving open the possibility that a greater variety of voices will be heard in policy debate. Mr Jock Munro, who I introduced earlier, continues to argue before various inquiries for the return to the days of wheat industry stabilisation policies (Australia. Productivity Commission 2010a) and to rail against the free market and the big players in news blogs (Sutton 2011). In his evidence to the 2010 Productivity Commission Inquiry into Wheat Marketing Arrangements he stated

> Just talking about the single desk, the core principles, we had a collective marketing power. It completely enrages me that people can talk about competition as a benefit to growers when all this system now does is has us competing against each other as weak individual growers. (Australia. Productivity Commission 2010a: 494)

Whereas previously he had the backing of AWB Limited and the Grains Council for his views, now he is one of many voices.

To return to Mahoney's typology, one of the reasons for the loss of support for AWB Limited was the mixed reaction to the revelations of the Cole Inquiry. While some supporters continued to defend the organisation's actions in Iraq, others could see that the institution had lost its way and the government lost patience. As Mahoney argues (2000: 525),

> The legitimacy underlying any given institution can be cast off and replaced when events bring about its forceful juxtaposition with an alternative, mutually incompatible conceptualization. Depending on the specific institution in question, the events that trigger such changes in subjective perceptions and thus declines in legitimacy may be linked to structural isomorphism with rationalized myths, declines in institutional efficacy or stability, or the introduction of new ideas by political leaders.

All of these factors appear to have been at play in the demise of collective wheat marketing in Australia.

Some Policy Lessons from the Death of Collective Wheat Marketing

The demise of collective wheat marketing in Australia contains some lessons for policy makers in two important areas. The first relates to the privatisation process which left the export monopoly in the hands of a private company, albeit one with an odd share structure. Second, AWB Limited's activities during Oil-For-Food provide a dramatic example of the limitations of economic sanctions as an instrument of international relations.

How Not to Privatise a Monopoly

With respect to the first issue, the group of countries which may learn from Australia's experience with the end of export wheat marketing specifically is an exclusive club of one – Canada. Lesson-drawing in public policy is common and as Rose (1991: 4) has noted

> it raises the possibility that policymakers can draw lessons that will help them deal better with their own problems. If the lesson is positive, a policy that works is transferred, with suitable adaptations. If it is negative, observers learn what not to do from watching the mistakes of others.

The structure of the entity which emerged from the privatisation of the Australian Wheat Board was a mistake. This judgement is not only made with the benefit of hindsight nor based on the unfortunate events surrounding the Oil-for-Food program. Concerns were raised before the privatisation was complete. There are several flawed components to the process that are worth noting. First the twin grower goals of grower control and the continuation of the export monopoly, while consistent with the values underpinning the old Wheat Board, were arguably incompatible with a private company structure. The two classes of shares that were introduced set up an apparent conflict of interest between maximising returns to growers, the company's customers, and maximising returns to shareholders. Granting an export monopoly to a private company with only the limited oversight provided by the Wheat Export Authority was also a problem. The monopoly did not have the unanimous support of wheatgrowers and it is likely that, if measured in terms of export earnings, the greater part of the industry would have preferred a free market.

The second flaw relates to the process of the privatisation. The nature of the closed policy debate meant that there were only three real participants in the policy discussion: the Department of Primary Industries and Energy, the Grains Council of Australia and the Wheat Board itself. The Department took a largely hands-off approach to the debate as a result of stated Ministerial preference that the industry come up with its own model for the Board restructure. The Grains Council was still closely attached in values terms to the collective marketing of the wheat crop and struggled to retain the key features of that system in the face of a strong push by the Wheat Board for privatisation. The debate should have included other key government departments and the policy community should have been open to the views of growers who did not necessarily subscribe to the GCA's values – and to other members of the grain value chain.

At this point it is worth a brief diversion to discuss another occasion on which the government backed away from an opportunity for a rigorous consideration of the wheat marketing arrangements. As noted in Chap. 5, the GCA was successful, in the context of the 1996 federal election campaign, in gaining a commitment from both sides of politics relating to the National Competition Policy. It persuaded both the incumbent, but soon to be opposition, Labor Party and the National Party that the *Wheat Marketing Act* as amended to reflect the (at the time undecided) restructure of the Wheat Board, should be pushed to the back of the queue of legislation to be reviewed under the National Competition Policy's legislative review program. Under the National Competition Policy an Agreement was signed between the Commonwealth, State and Territory Governments which *inter alia* required the jurisdictions to develop 'a timetable by June 1996 for the review and, where appropriate, reform of all existing legislation which restricts competition by the year 2000' (National Competition Council 1998: 40). These reviews were to remove anti-competitive behaviour and were driven by a clear set of principles, including:

> 5(1) The guiding principle is that legislation (including Acts, enactments, ordinances or regulations) should not restrict competitions unless it can be demonstrated that
>
> (a) the benefits of the restriction to the community as a whole outweigh the costs; and
> (b) the objectives of the legislation can only be achieved by restricting competition. (National Competition Council 1998: 19)

While the Grains Council and many in the industry were confident that the monopoly arrangements were delivering net benefits to wheat growers, they were nervous that they would fail the broader 'community interest' test.

The National Competition Policy review process, although seen as a major threat during the course of the privatisation debate, proved to be something of a damp squib. The Government chose to apply a light touch to the National Competition Policy review of the wheat marketing legislation in 2000 and it was a lost opportunity for detailed scrutiny of the arrangements. The reasoning behind the failure to have the Productivity Commission undertake the NCP review of the *Wheat Marketing Act* was inadvertently revealed by a National Party Senator in 2007 when he mentioned, not once but three times, that his party's political opponents, the Labor Party's policy of having the export monopoly reviewed by the Productivity Commission would be 'the kiss of death for any single desk' (Boswell 2007: 30).

The Productivity Commission made a submission to the review in which they focused on the advantages and disadvantages of the export monopoly. The submission concluded that 'it is unlikely that the current wheat export marketing monopoly generates net benefits for Australia or, indeed, wheat producers themselves' (Australia. Productivity Commission 2000: 12). The Review team appeared to agree, arguing that 'Regarding the public benefits test, the Committee was not presented with, nor could it find, clear, credible, and unambiguous evidence that the current arrangements for the marketing of export wheat are of net benefit to the Australian community' (Irving et al. 2000: 6). However, rather than calling for the removal of the monopoly, the Review recommended that

> **the 'single desk' be retained until the scheduled review in 2004 by the Wheat Export Authority (WEA) of AWBI's operation of the 'single desk'. However, the main purpose and implementation of this scheduled review should be changed so that it provides one final opportunity for a compelling case to be compiled that the 'single desk' delivers a net benefit to the Australian community.** (Irving et al. 2000: 8 – emphasis in original)

Although the National Competition Council found that the review of the wheat marketing legislation 'was open, independent and rigorous', it concluded that 'the Commonwealth Government had not met its [competition principles agreement] clause 4 and 5 obligations arising from the Wheat Marketing Act' (National Competition Council 2003: 8). These unmet obligations related to structural reform of public monopolies and legislation review and reform.

The lesson from both the privatisation process itself and the subsequent relatively limp legislative review is that there are risks associated with leaving policy in the hands of a limited group of interests. A policy as significant and an asset as valuable as a monopoly should be subject to greater and wider levels of public scrutiny. In Australia, it took the Oil-for-Food scandal to focus attention on these arrangements. It is possible that the scandal simply accelerated the inevitable. As explained earlier, the US has its sights on statutory marketing arrangements in agriculture in the Doha Round of multilateral trade negotiations and the latest negotiating document specifically targets export monopolies. A lesson to be learnt, if there is one, from the Australian Wheat Board privatisation is to have an open and transparent process of change which engages all of the affected players and which puts all options on the table.

The Problem of Sanctions Implementation

The second policy lesson to be drawn from this case study relates to the enforcement at the domestic level of economic sanctions agreed to in the international arena. One area in which the Australian Government could perhaps be found culpable with regard to the Oil-for-Food scandal relates to its failure to set up adequate national mechanisms to ensure that Australian nationals complied with the sanctions regime. While nation states agree to the imposition of sanctions it falls to private individuals and companies to implement them, and often at great cost to themselves. Without an effective domestic legislative framework to ensure compliance, companies can bypass the sanctions without breaking any law. This was one of the fundamental problems faced by the Government in responding to the revelations of the Cole Inquiry. While it was generally agreed that the activities of AWB Limited were unethical, they were not illegal under Australian law and they were arguably not corrupt (Botterill 2011). The most the authorities could find against some AWB Limited office bearers was that they were in breach of elements of Australia's corporations law. The complexity of the Oil-for-Food program, which was essentially an elaborate set of exceptions from the sanctions, left the scheme open for corruption at many levels as well as providing opportunities for the kickback arrangements of which AWB Limited was part. The resulting scandal reinforces Doxey's (1996: 106) observation that 'Far-reaching exemptions undermine the effectiveness of the sanctions.'

The two policy lessons are linked. The Australian Government became entangled in the scandal surrounding AWB Limited because of the export monopoly. The wheat marketing legislation granted an effective monopoly to the company and the government agency Wheat Exports Australia had a statutory role in its oversight. Both of these factors meant that any issues with AWB Limited could not be dismissed as a private company behaving badly which could be condemned by government. The Government was by implication involved in the scandal. It did not have the legal power to demand information from AWB Limited about its activities once alerted to suspicions about possible kickbacks and there was no regulatory framework in place to ensure enforcement of the sanctions under domestic law.

It was also considered lax in its oversight of the sanctions regime because officers within the Department of Foreign Affairs and Trade (DFAT) were accused of inadequate scrutiny of the wheat export contracts sent to them by AWB Limited. There are two ameliorating factors that need to be borne in mind when considering the role of the DFAT officers in this affair. The first relates to the fact that the privatisation of the Australian Wheat Board greatly diminished government access to expertise about the international grain market. As the holder of the export monopoly the statutory Wheat Board was also the effective monopoly holder of detailed information on its export performance and the international wheat trade more generally (Ginns 2007: 127). The coincidence in timing of the privatisation process and the implementation of Oil-for-Food meant that officers within the Department lost their access to the expertise of the Wheat Board at precisely the time they needed it in order to make sound judgements about the wheat prices in the contracts they were scrutinising.

There is another, more personal element to this story. The personnel at AWB Limited who dealt with DFAT were in large part the same individuals on whom the Department had relied for market intelligence prior to the privatisation. Mulgan suggests that DFAT officers may have continued to have faith that AWB Limited's employees with whom they were dealing and who only recently had been public servants and colleagues, would remain 'bound by public service procedures and expectations of integrity' (Mulgan 2009: 341). I argue that the trust issue would occur at a deeper interpersonal level. As Granovetter (1985: 492) notes, 'The trust engendered by personal relations presents, by its very existence, enhanced opportunity for malfeasance'. While his study is focused on the role of networks of interpersonal relations in determining behaviour in an economic context, for example interactions between firms, Granovetter's analysis is apposite in the context of the AWB Limited case. The privatisation process should have alerted the DFAT officers that their interlocutors at AWB Limited were now operating within a different organisational context; but it is not clear that officers at DFAT would necessarily have known the progress of that privatisation given the two stage nature of the process and the low level of public scrutiny. In addition to their trust in AWB Limited to be truthful, DFAT showed a faith in UN processes for examining the wheat contracts which, with the benefit of hindsight, also appears naive. As the Cole Inquiry noted of the disguising of the trucking fee (2006: xxv), 'The contracts submitted with the associated documents to DFAT and the United Nations did not record or reflect the true arrangements between AWB and the IGB [Iraqi Grains Board]'. The report further notes that 'These matters were deliberately and dishonestly concealed from DFAT and the United Nations' (Cole 2006: xxv–xxvi).

Collective Wheat Marketing in Australia

For 60 years, the Australian wheat industry was characterised by a collective ethos, embodied first in the *Wheat Stabilization Act 1948* and later in the *Wheat Marketing Act 1989*. The collective arrangements were given organisational form through the statutory Australian Wheat Board and its privatised successor, AWB Limited. Over the course of its history the institution of collective wheat marketing came under pressure from the forces of economic deregulation and reform, however, it managed to preserve its two core values of the export monopoly and grower control until 2008. The success of the institution arose from a combination of factors. First, it had the support of the peak grains industry body, the Grains Council of Australia; support which the Wheat Board and AWB Limited buttressed by providing increasingly large amounts of financial support to the organisation. Second, it was operating in an area of economic policy, rural policy, which is treated with benign indifference by the general Australian population and the mainstream media and was therefore left in the hands of an exclusive policy community to develop. Third, the industry played the political game effectively, particularly with obtaining the delay to the legislative review of the wheat marketing legislation. It could also always count on the National Party to defend the arrangements.

The institution's demise in 2008 arose from the actions of a small group of AWB Limited employees during the Oil-for-Food scandal. However, the environment within which the company was operating had changed. Domestically, the Cole Inquiry effectively expanded the policy community surrounding wheat industry policy by drawing the attention of other groups, such as opponents of the single desk and shareholder activists, to the nature of the company and its operations. The international environment had also changed with the general acceptance of the operation of State Trading Enterprises which had characterised the early years of the GATT replaced by an explicit targeting of agricultural STEs in the Doha Round of multilateral trade negotiations. Both the Labor and Liberal Parties showed a general inclination to end the export monopoly for wheat. I have sought to demonstrate in this book, though, that although the Oil-for-Food scandal was the proximate cause of the demise of collective wheat marketing, the seeds had been sown as a result of the earlier policy change when the government deregulated the domestic wheat market resulting in a change of marketing strategy by the Wheat Board.

From a historical institutionalist perspective, the wheat marketing arrangements resulted from a lengthy debate in which the values of collective agrarianism won over the free market values of the larger producers and the wheat merchants. Once institutionalised as the Australian Wheat Board, the arrangements were protected against changing economic circumstances through strategic, incremental changes. These adjustments protected the core values of the export monopoly and grower control from serious challenge. In the mid-1990s the industry took the initiative with a discussion about the restructuring of the statutory marketing body. Many of the rationales for the export monopoly, such as the export subsidy wars between the US and the EU during the 1980s, had disappeared and the industry's own modelling undertaken in the context of the Strategic Planning Unit process indicated that the benefit to growers of the monopoly was marginal at best. However, attachment to the export monopoly was an article of faith for many growers, irrespective of its instrumental value, and it is on this basis that I have argued that the monopoly was a value. The collective wheat marketing arrangements were firmly anchored in agrarian ideas about agriculture and a profound suspicion of the 'middleman'. And the industry has a long memory. The dire conditions of the 1930s have not been forgotten in the early twenty-first century as was evidenced during the hearings held by the Productivity Commission in 2010 when some growers continued to call for re-regulation of wheat marketing.

The story of collective wheat marketing in Australia provides an illustration of the value of the historical institutionalist approach to understanding policy evolution and change and it adds a further dimension in considering the death of an institution. It is a truism that 'history matters', however it is too often forgotten in policy analysis and commentary. It is hoped that this small contribution will assist in a more nuanced understanding of the history of the Australian Wheat Board and AWB Limited by filling some gaps in general knowledge and understanding about the industry's recent institutional history.

References

AAP (2011) Cattle exports banned to Indonesia. The Australian, 8 June 2011

Australia. Productivity Commission (2000) Productivity Commission Submission to the National Competition Policy Review of the *Wheat Marketing Act 1989*. July 2000. http://www.pc.gov.au/research/submission/wheat1. Accessed October 2010

Australia. Productivity Commission (2010a) Draft report on wheat export marketing arrangements. Transcript of Proceedings at Sydney on Tuesday 11 May 2010. http://www.pc.gov.au/__data/assets/pdf_file/0011/98480/20100511-sydney.pdf

Australia. Productivity Commission (2010b) Wheat export marketing arrangements. Productivity Commission Inquiry Report, No. 51, Canberra, 1 July 2010

Beyer J (2010) The same or not the same—on the variety of mechanisms of path dependence. Int J Soc Sci 5(1):1–11

Boswell R, Senator (2007) Wheat Marketing Amendment Bill 2007: Second Reading Debate. Senate Hansard, 21 June 2007

Botterill LC (2006) Soap operas, cenotaphs and sacred cows: countrymindedness and rural policy debate in Australia. Public Policy 1(1):23–36

Botterill L (2009) The role of agrarian sentiment in Australian rural policy. In: Merlan F, Raftery D (eds) Tracking rural change: community, policy and technology in Australia, New Zealand and Europe. ANU E Press, Canberra

Botterill LC (2011) Circumventing sanctions against Iraq in the Oil for Food program. In: Graycar A, Smith R (eds) Global handbook on research and practice in corruption. Edward Elgar

Brunsson N, Olsen JP (1993) The reforming organization. Routledge, London

Campbell JL (1998) Institutional analysis and the role of ideas in political economy. Theor Soc 27:377–409

Cole Inquiry (2006) Inquiry into certain Australian companies in relation to the UN Oil-for-Food programme. Transcript. http://www.offi.gov.au/agd/WWW/unoilforfoodinquiry.nsf/Page/Transcripts. Accessed 6 June 2011

Cole TR, the Honourable AO RFD QC (2006) Report of the inquiry into certain Australian companies in relation to the UN Oil-for-Food programme. Volume 1: Summary, recommendations and background, Commonwealth of Australia, Canberra

Daugbjerg C, Botterill L (2011) Ethical food standard schemes and global trade: challenging the WTO? 85th Agricultural Economics Society Conference, Warwick, UK, 18–20 April

Doern GB, Phidd RW (1983) Canadian public policy: ideas, structure, process. Methuen, Toronto

Dowding K (1994) The compatibility of behaviouralism, rational choice and 'new institutionalism'. J Theor Polit 6(1):105–117

Doxey MP (1996) International sanctions in contemporary perspective, 2nd edn. Macmillan Press Ltd, Basingstoke

Etzioni A (1988) The moral dimension: toward a new economics. The Free Press, New York

Ginns D (2007) Against the grain: the AWB scandal and why it happened—by Stephen Bartos. Aust J Publ Admin 66:126–127

Granovetter M (1985) Economic action and social structure: the problem of embeddedness. Am J Sociol 91(3):481–510

Irving M, Arney J, Lindner B (2000) National Competition Policy Review of the Wheat Marketing Act 1989. Commonwealth of Australia, Canberra, December 2000

Lasswell HD (1951) The policy orientation. In: Lerner D, Lasswell HD (eds) The policy sciences. Stanford University Press, Stanford

Mahoney J (2000) Path dependence in historical sociology. Theor Soc 29:507–548

March JG, Olsen JP (1984) The new institutionalism: organizational factors in political life. Am Polit Sci Rev 78(3):734–749

Marsh D, Rhodes RAW (1992) Policy communities and issue networks: beyond typology. In: Marsh D, Rhodes RAW (eds) Policy networks in British government. Clarendon Press, Oxford

Merton RK (1936) The unintended consequences of purposive action. Am Sociol Rev 1(6):894–904

Mulgan R (2009) AWB and oil for food: some issues of accountability. In: Farrall J, Rubenstein K (eds) Sanctions, accountability and governance in a globalised world. Cambridge University Press, Cambridge

National Competition Council (1998) Compendium of National Competition Policy Agreements. Commonwealth of Australia, Canberra. http://ncp.ncc.gov.au/docs/PIAg-001.pdf

National Competition Council (2003) Assessment of governments' progress in implementing the National Competition Policy and related reforms: 2003 - volume two: legislation review and reform. AusInfo, Canberra

Peters BG, Pierre J (1998) Institutions and time: problems of conceptualization and explanation. J Publ Admin Res Theor 8(4):565–583

Pierson P (2004) Politics in time: history, institutions, and social analysis. Princeton University Press, Princeton

Rayner J (2009) Understanding policy change as a historical problem. J Comp Policy Anal: Res Pract 11(1):83–96

Rose R (1991) What is lesson-drawing? J Public Policy 11(1):3–30

Sabatier P (1988) An advocacy coalition framework of policy change and the role of policy-oriented learning therein. Policy Sci 21:129–168

Sutton M (2011) Cargill raises stakes for SA grain chain. Stock Journal, 12 May 2011

Vaughan D (1999) The dark side of organizations: mistake, misconduct, and disaster. Annu Rev Sociol 25:271–305

Index

A
ABB Ltd. *See* Australian Barley Board Ltd.
Accreditation scheme, 111
Advocacy coalitions, 6, 86
After sales service fee, 95
Agents, 3, 36, 67, 95, 129
Agrarian collectivism, 34, 48, 51, 129, 130, 131, 133, 149
Agrarian ideals, 1, 11
Agrarianism, 11, 12, 13, 131, 139
Agri-politics, 61, 62, 117
Agrium, 113
Alia for General Transportation and Trade (Alia), 95
Andersen, A., 97
Anderson, J., 74, 79, 83
Annan, K., 94
ASW. *See* Australian Standard White
Australian Barley Board (ABB) Ltd., 60
Australian Labor Party, 80
Australian Shareholders Association, 112
Australian Standard White (ASW), 57
Australian Stock Exchange, 82
Australian Wheat Board (AWB) Limited, 2, 4, 5, 10, 15, 16, 41–46, 58, 59, 61, 67–71, 73–85, 87, 91, 94–105, 109, 112–119, 123, 127–133, 135, 137–140
Australian Wheatgrowers Federation (AWF), 5, 40–43, 48, 51, 52, 58, 61, 65, 70, 117–119, 133
Australia-US Free Trade Agreement, 111–112
AWB Limited. *See* Australian Wheat Board Limited
AWF. *See* Australian Wheatgrowers Federation

B
BHP Petroleum, 98
Bulk handling authorities, 60

C
Cairns Group, 55
Canada, 1, 27, 45, 46, 48, 114, 116, 135
Canadian Wheat Board (CWB), 1, 44–46, 58, 116
Canadian Wheat Board Act, 46
Cargill, Incorporated, 113
Chifley, J.B., 30, 42, 43
Coalition, 6, 10–11, 16, 36–40, 53, 55, 70, 71, 79, 80, 83, 84, 86, 92, 105, 107, 108, 110
Coalition of the willing, 105
Cole Inquiry. *See* Inquiry into certain Australian companies in relation to the UN Oil-for-Food Programme
Collective marketing, 1–2, 6, 14, 15, 16, 67, 68, 71, 86
Collective wheat marketing, 1, 2, 6, 10, 15, 33–48, 52, 55, 65, 67, 69, 86, 87, 134–140
Collectivism, 1, 15, 34, 48, 51, 130, 133
Commodity councils, 52, 63, 64
Competition Council, 72, 79
Complementary institutions, 5, 52, 133
Compulsory acquisition, 62, 72
Compulsory pooling, 35, 36, 37, 39, 40, 41, 42
Co-operatives, 13, 35, 38, 43, 45, 60, 73, 74, 80, 111
Corporatisation, 77, 79, 83
Country mindedness, 13
Country Party, 10, 39, 53, 55, 84

L.C. Botterill, *Wheat Marketing in Transition: The Transformation of the Australian Wheat Board*, Environment & Policy 53, DOI 10.1007/978-94-007-2804-2,

Critical junctures, 4, 5, 14–15, 67, 68
CWB. *See* Canadian Wheat Board

D
Deep core values, 6, 86
Department of Finance, Australia, 84
Department of Foreign Affairs and Trade, Australia, 16, 95, 96, 111, 127–128, 138
Department of Primary Industries and Energy, Australia, 64, 73, 76, 79, 80, 83, 84, 135
Depression, 27, 43–45
Deregulation, 6, 15, 21, 51–65, 67–87, 109, 110, 113, 114, 124, 131, 139
Doha Development Round, 46, 111, 115, 116, 137, 139
Domestic deregulation, 67–87
Drought, 12, 23–25, 45, 64

E
Economic deregulation, 69, 139
Economic sanctions, 91, 99, 102, 129, 130, 134, 137
Escrow account, 93–95, 98, 100
European Union, 27, 115
Exogenous shock, 4, 67, 128
Export monopoly, 2, 4–6, 10, 11, 15, 16, 44, 57, 59, 60, 65, 67, 68, 71, 73–79, 81–87, 91–105, 109, 110, 113, 114, 117, 123–124, 130, 135–140
Exports, 14, 22, 26–31, 82, 83, 91–102, 109–111, 115, 127, 137

F
Fair Average Quality (FAQ), 56, 57, 65, 131
Family farms, 10–12
FAQ. *See* Fair Average Quality
Farm incomes, 14, 24, 31
Firewalls, 7, 8, 64
First World War, 24, 30, 35, 39, 43
Flugge, T., 78, 79, 118–119
Free market liberalism, 34, 65

G
Ginns, D., 119, 120, 122–123, 138
GMP. *See* Guaranteed minimum price
Government policy, 14, 15, 29, 65, 76, 83–85, 111
Government underwriting, 21, 58, 69, 74, 117
GPA. *See* Grain Producers Australia
Grainco, 60, 74
Grain cooperatives, 74
Graincorp, 60, 70, 113, 123, 127
Graingrowers, 5, 34, 52, 74, 85, 100, 120, 127, 130
Grain Producers Australia (GPA), 121–123
Grains 2000, 73
Grains Council of Australia, 67–70, 73–75, 77–80, 84, 85
Grains Research and Development Corporation, 73, 120, 122, 123
Grains Week, 76, 80, 81, 83, 101
Grain traders, 34, 59, 71, 86, 101, 102, 130–131
Graziers, 29, 63
Grower control, 1, 43, 76–80, 82, 85–87,109, 112–114, 123–124, 135, 139
Grower meetings, 74, 77–79, 101, 112, 118
Grower organisations, 2, 40, 43, 44, 118, 120
Guaranteed minimum price (GMP), 21, 36, 58, 59, 69, 71
Guaranteed price, 35–37, 41, 42

H
Hilmer Report. *See* National Competition Policy
Historical institutionalism, 3–6, 68, 102, 107, 129, 133
Historical institutionalist, 3, 4, 6, 8, 48, 65, 67, 128, 129, 133, 139, 140
Home consumption price, 21, 35, 37, 38, 41, 42, 44
Home price scheme, 30
Humanitarian imports, 93, 100
Humanitarian needs, 93
Hussein, S., 16, 91, 93–95, 99, 100, 110

I
IAC. *See* Industries Assistance Commission
Income stability, 31, 33, 46
Incrementalism, 7, 8
Industries Assistance Commission (IAC), 52–55, 57–62, 65, 67, 69, 71, 85, 118
Industry assistance, 39, 53, 55
Industry Commission, 62
Inquiry into certain Australian companies in relation to the UN Oil-for-Food Programme, 16, 94
Institutional change, 4, 5, 67–71
Institutional layering, 68
Institutional reproduction, 3, 51, 102–104, 132
Institutional survival, 3, 5, 6, 103, 107

Institutional transformation, 4, 68
International grain markets, 16, 138
International Wheat Agreement (IWA), 14, 46–48, 114–115
International Wheat Council, 46
International wheat market, 27, 46
Iran, 27–28
Iraq, 2, 10, 15, 27, 91–102, 104, 105, 108, 110, 112, 113, 119, 127, 128, 130, 133, 134
Iraqi government, 92–95, 98, 99
Iraqi wheat debt, 92, 127
IWA. *See* International Wheat Agreement

K
Keating, P., 72
Kickbacks, 2, 15, 16, 93–95, 97, 101–105, 110, 130, 133, 137, 138
Kuwait, 91–92

L
Labor Party. *See* Australian Labor Party
Land reforms, 29, 63
Legitimacy, 5, 7, 16, 68, 104, 133, 134
Levies, 59, 61, 120, 122, 123
Liberal-National Party coalition, 10, 11, 80, 108
Liberal Party of Australia, 10

M
Macfarlane, I., 76, 78, 79
Marketers, 60, 70, 71, 107, 130, 132–133
Market liberalism, 34, 52, 65, 114
Media, 8–9, 13, 15, 104, 108, 112–114, 127–128, 139
Merchants, 35, 36, 37, 39, 70, 93, 139
Microeconomic reform, 72
Middle East, 27, 93
Middle men, 1, 16, 36, 86, 103

N
National Competition Policy, 62, 72, 76, 77, 79, 80, 83, 84
National Farmers Federation (NFF), 52, 55, 63–65, 85, 120, 123
National Grain Marketing Strategic Planning Unit, 75
National Grains Australia, 122
National Party of Australia, 71
Newco, 73–75, 77, 78
New institutionalisms, 3, 129–130
New South Wales, 24, 26, 29–30, 40–41, 69, 85, 122, 123
New South Wales Farmers and Settlers' Association, 40
NFF. *See* National Farmers Federation

O
Oil-for-Food program, 87, 91–98
Oil smuggling, 93
Orderly marketing, 51–65, 71, 86
Organizational culture, 103

P
Path dependence, 4, 5, 67, 103
Policy communities, 3, 8, 9, 16, 34, 129, 134
Policy cycling, 7
Policy networks, 3, 8, 9, 34
Pooling, 4, 35–37, 39–42, 44, 45, 58, 60, 62, 65, 80, 84, 85, 104, 116, 119
Powers of acquisition, 69
Price stability, 47
Primary Producers' Association of Western Australia, 35, 40, 41, 42
Privatisation, 13, 55, 60, 62, 67–87, 101, 102, 112, 113, 117–119, 128, 130–132, 134–138
Productivity Commission, 22, 25, 28, 111, 112, 114, 123, 124, 127, 134, 136, 140
Public policy, 2, 3, 51, 64, 128, 135

Q
Queensland, 24, 26, 38, 43, 60, 69, 74, 85, 117
Queensland Graingrowers Association, 74

R
Rational choice, 3, 129
Royal Commission into Grain Transport, Storage and Handling, 60, 65
Royal Commission on the Wheat, Flour and Bread Industries, 34, 57
Rural exports, 14, 22
Rural policy, 8–11, 13, 14, 29–31, 43, 51–55, 63–65
Rural Policy Green Paper, 1974, 52, 54, 55, 57
Rural Reconstruction Commission, 1943, 30

S

Sanctions implementation, 16, 98–101, 137–138
Second World War, 27, 30, 39, 41, 43
Shareholders, 81, 82, 108, 112–113, 127, 135, 139
Share structure, 80
Single desk. *See* Export monopoly
SMAs. *See* Statutory Marketing Authorities
Soldier settlement schemes, 30, 63
South Australia, 24, 26, 35, 40, 60
Squatters, 29–30
Stabilisation, 21, 31, 33, 39, 42, 43, 45, 51, 54, 56–58, 65, 114, 134
State trading, 46, 47, 115–117, 124, 139
State trading enterprises (STEs), 46, 47, 115–117, 139
Statutory marketing, 54, 61–62, 65, 85
Statutory Marketing Authorities (SMAs), 61–62, 85
STEs. *See* State trading enterprises
Strategic planning, 15, 75, 84
Strategic Planning Units, 123, 140
Strategies of reproduction, 3–5, 8, 51, 102, 103, 114, 128
Structural adjustment, 3, 23, 44, 64

T

Tariff Board, 52–53
Tasmania, 24–26, 38
Tigris, 98, 105
Traders, 34, 41, 47, 48, 57, 59, 70, 71, 85, 86, 101–103, 107, 113, 130
Transportation fee, 94–96, 98, 101

U

UAP. *See* United Australia Party
Unintended consequences, 3, 6, 102, 128
United Australia Party (UAP), 36, 37, 39, 41
United Grain Growers Limited, 82
United Kingdom, 9, 10, 27, 95
United Nations, 2, 15, 16, 91–98, 100, 101, 103, 105, 110, 113, 138
United Nations Independent Inquiry Commission, 94
United Nations Security Council, 91–93
UN Office of the Iraq Program, 95, 96
Uruguay Round (of multilateral trade negotiations), 51, 55, 75–77, 115, 117
US Wheat Associates, 2, 116

V

Value conflicts, 7, 8, 64
Values, 1–10, 12–16, 22, 28, 33, 34, 42–44, 48, 51, 52, 55, 56, 59, 60, 64, 65, 67, 68, 71, 74–77, 82, 85, 86, 95, 98, 102–104, 107, 108, 113–117, 119, 123, 127–135, 139, 140
VicGrain, 60
Victoria, 24, 26, 30, 35, 40–42, 60, 108
Victorian Wheat and Woolgrowers Association, 41
Volcker Inquiry/Report. *See* United Nations Independent Inquiry Commission

W

WEA. *See* Wheat Export Authority
Western Australia, 10, 24–26, 28, 35, 36, 40–42, 56, 60, 84–85, 122, 123, 131
Wheat belt, 24–26, 57, 77
Wheat Board. *See* Australian Wheat Board
Wheat Bounty Act 1934, 37
Wheat Export Authority (WEA), 82, 109, 111, 135, 136
Wheat Export Marketing Act 2008, 111
Wheat Export Marketing Consultative Committee, 109
Wheat Exports Australia, 2, 4–6, 14, 26–28, 36, 57, 91–102, 111, 115, 137
Wheatgrowers, 2, 5, 35, 37, 39–44, 48, 51, 52, 57, 58, 61, 69, 71, 81, 82, 85, 103, 112, 114, 117–119, 133, 135, 136
Wheat Growers Relief Act (No 2) 1934, 37
Wheatgrowers' Union of New South Wales, 40
Wheat Industry Assistance Act 1948, 39
Wheat Industry Fund, 69, 73, 75–77, 79, 81, 82
Wheat industry policy, 14, 21–31
Wheat Marketing Act 1984, 58, 118
Wheat Marketing Act 1989, 4, 67, 80–82, 118, 139
Wheat Marketing Bill 1930, 35
Wheat quotas, 27, 56
Wheat Stabilization Act 1948, 4, 33, 34, 41, 42, 51, 139
Wheat trade, 16, 33
White Paper, 1986, 61, 62
Wool, 22, 23, 29, 31, 41
World Trade Organization (WTO), 46, 115–117
WTO. *See* World Trade Organization